AF333826

Oliver Kramer

# Self-Adaptive Heuristics for Evolutionary Computation

# Studies in Computational Intelligence, Volume 147

**Editor-in-Chief**

Prof. Janusz Kacprzyk
Systems Research Institute
Polish Academy of Sciences
ul. Newelska 6
01-447 Warsaw
Poland
*E-mail:* kacprzyk@ibspan.waw.pl

Further volumes of this series can be found on our homepage:
springer.com

Vol. 126. Oleg Okun and Giorgio Valentini (Eds.)
*Supervised and Unsupervised Ensemble Methods
and their Applications*, 2008
ISBN 978-3-540-78980-2

Vol. 127. Régie Gras, Einoshin Suzuki, Fabrice Guillet
and Filippo Spagnolo (Eds.)
*Statistical Implicative Analysis*, 2008
ISBN 978-3-540-78982-6

Vol. 128. Fatos Xhafa and Ajith Abraham (Eds.)
*Metaheuristics for Scheduling in Industrial and Manufacturing
Applications*, 2008
ISBN 978-3-540-78984-0

Vol. 129. Natalio Krasnogor, Giuseppe Nicosia, Mario Pavone
and David Pelta (Eds.)
*Nature Inspired Cooperative Strategies for Optimization
(NICSO 2007)*, 2008
ISBN 978-3-540-78986-4

Vol. 130. Richi Nayak, Nikhil Ichalkaranje
and Lakhmi C. Jain (Eds.)
*Evolution of the Web in Artificial Intelligence Environments*,
2008
ISBN 978-3-540-79139-3

Vol. 131. Roger Lee and Haeng-Kon Kim (Eds.)
*Computer and Information Science*, 2008
ISBN 978-3-540-79186-7

Vol. 132. Danil Prokhorov (Ed.)
*Computational Intelligence in Automotive Applications*, 2008
ISBN 978-3-540-79256-7

Vol. 133. Manuel Graña and Richard J. Duro (Eds.)
*Computational Intelligence for Remote Sensing*, 2008
ISBN 978-3-540-79352-6

Vol. 134. Ngoc Thanh Nguyen and Radoslaw Katarzyniak (Eds.)
*New Challenges in Applied Intelligence Technologies*, 2008
ISBN 978-3-540-79354-0

Vol. 135. Hsinchun Chen and Christopher C. Yang (Eds.)
*Intelligence and Security Informatics*, 2008
ISBN 978-3-540-69207-2

Vol. 136. Carlos Cotta, Marc Sevaux
and Kenneth Sörensen (Eds.)
*Adaptive and Multilevel Metaheuristics*, 2008
ISBN 978-3-540-79437-0

Vol. 137. Lakhmi C. Jain, Mika Sato-Ilic, Maria Virvou,
George A. Tsihrintzis, Valentina Emilia Balas
and Canicious Abeynayake (Eds.)
*Computational Intelligence Paradigms*, 2008
ISBN 978-3-540-79473-8

Vol. 138. Bruno Apolloni, Witold Pedrycz, Simone Bassis
and Dario Malchiodi
*The Puzzle of Granular Computing*, 2008
ISBN 978-3-540-79863-7

Vol. 139. Jan Drugowitsch
*Design and Analysis of Learning Classifier Systems*, 2008
ISBN 978-3-540-79865-1

Vol. 140. Nadia Magnenat-Thalmann, Lakhmi C. Jain
and N. Ichalkaranje (Eds.)
*New Advances in Virtual Humans*, 2008
ISBN 978-3-540-79867-5

Vol. 141. Christa Sommerer, Lakhmi C. Jain
and Laurent Mignonneau (Eds.)
*The Art and Science of Interface and Interaction Design (Vol. 1)*,
2008
ISBN 978-3-540-79869-9

Vol. 142. George A. Tsihrintzis, Maria Virvou, Robert J. Howlett
and Lakhmi C. Jain (Eds.)
*New Directions in Intelligent Interactive Multimedia*, 2008
ISBN 978-3-540-68126-7

Vol. 143. Uday K. Chakraborty (Ed.)
*Advances in Differential Evolution*, 2008
ISBN 978-3-540-68827-3

Vol. 144. Andreas Fink and Franz Rothlauf (Eds.)
*Advances in Computational Intelligence in Transport, Logistics,
and Supply Chain Management*, 2008
ISBN 978-3-540-69024-5

Vol. 145. Mikhail Ju. Moshkov, Marcin Piliszczuk
and Beata Zielosko
*Partial Covers, Reducts and Decision Rules in Rough Sets*, 2008
ISBN 978-3-540-69027-6

Vol. 146. Fatos Xhafa and Ajith Abraham (Eds.)
*Metaheuristics for Scheduling in Distributed Computing
Environments*, 2008
ISBN 978-3-540-69260-7

Vol. 147. Oliver Kramer
*Self-Adaptive Heuristics for Evolutionary Computation*, 2008
ISBN 978-3-540-69280-5

Oliver Kramer

# Self-Adaptive Heuristics for Evolutionary Computation

 Springer

Oliver Kramer
Computational Intelligence Group (LS 11)
Department of Computer Science
Dortmund University of Technology
44221 Dortmund
Germany
E-mail: oliver.kramer@tu-dortmund.de

D466, Dissertation in Computer Science submitted to the Faculty of Computer Science, Electrical Engineering and Mathematics, University of Paderborn in partial fulfillment of the requirements for the degree of doctor rerum naturalium (Dr. rer. nat.).

ISBN 978-3-540-69280-5                    e-ISBN 978-3-540-69281-2

DOI 10.1007/978-3-540-69281-2

Studies in Computational Intelligence        ISSN 1860949X

Library of Congress Control Number: 2008928449



*Typeset* & *Cover Design:* Scientific Publishing Services Pvt. Ltd., Chennai, India.

Printed in acid-free paper
9 8 7 6 5 4 3 2 1
springer.com

# Abstract

Evolutionary algorithms are biologically inspired meta-heuristics. They perform randomized search with the help of genetic operators. The success of evolutionary search is often a result of proper parameter settings and knowledge, the practitioner integrates into the heuristics. An extended taxonomy for parameter setting techniques is proposed. Self-adaptation is an implicit parameter adaptation technique enabling the evolutionary search to control the strategy parameters automatically by evolution.

Evolution strategies exhibit plenty of mutation operators from isotropic Gaussian mutation to the derandomized covariance matrix adaptation. The biased mutation operator proposed in this book is a self-adaptive operator biasing mutations into beneficial directions. Our experimental analysis reveals that biased mutation is appropriate for search at the boundary of constrained search spaces. Another successful example for self-adaptive mutation is self-adaptive inversion mutation proposed in this work. Its strategy parameters determine the number of successive inversion mutations, i.e. swaps of edges in graphs. Experimental analysis and statistical tests show that the operator is able to speed up the optimization, in particular at the beginning of the search. A convergence analysis shows that self-adaptive inversion mutation converges to the optimum with probability one after a finite number of iterations.

Crossover is the second main operator of evolutionary search. The book proposes self-adaptive enhancements for structural crossover properties like crossover points. The experimental analysis reveals that the link between strategy parameters and fitness is weak for the proposed properties. At the end, the work concentrates on self-adaptation of step sizes in constrained search domains. It may result in premature fitness stagnation. We prove premature convergence for a $(1+1)$-EA under simplified conditions. Adaptive constraint handling heuristic are introduced that offer useful means to overcome premature stagnation.

# Acknowledgments

It is a pleasure to gratefully acknowledge the help and support of many people who made this work possible. First of all, I would like to thank my advisor, Professor Hans Kleine Bning for his support and encouragement at the University of Paderborn. I also thank Professor Gnter Rudolph for useful discussions and support at the final stage of my work at the Technical University of Dortmund. Furthermore, my grateful thanks go to Professor Hans-Paul Schwefel who guided me to the field of evolutionary computation for very fruitful discussions and useful contributions during our cooperative work on constraint handling. I would like to thank Professor Franz Josef Rammig and Professor Thomas Bck for their interest in my work.

Many thanks for all the interesting scientific and private discussions go to the former and current members of the working group *Knowledge Based Systems* at the University of Paderborn. In particular, I would like to thank Professor Benno Stein, Dr. Theodor Lettmann, Dr. Chuan-Kang Ting, Dr. Andreas Goebels, Steffen Priesterjahn, Uwe Bubeck, Alexander Weimer, Dr. Klaus Brinker, Natalia Akchurina, Isabela Anciutti, Heinrich Balzer, Dr. Elina Hotman and Christina Meyer. The same holds for the members of the *Computational Intelligence* group of the *Algorithm Engineering* chair at the Technical University of Dortmund, in particular Nicola Beume, Boris Naujoks, Mike Preuss, Hoi-Ming Wong and Dr. Udo Feldkamp.

I point out the fruitful collaboration and constructive contributions during my supervision of master and bachelor theses, especially of Patrick Koch, Stephan Brgger, Dejan Lazovic, Melanie Kirchner and Bartek Gloger. I owe special thanks for technical and organizational support to the International Graduate School Team in Paderborn, in particular Dr. Eckhard Steffen and Astrid Canisius. In this context I also thank Simone Auinger, Gundel Jankord, Patrizia Hfer, Gerd Brakhane and Ulrich Hermes.

I would like to express my deepest gratitude to my girlfriend Janina Baucks for her love and support at the final stage of my work. Last but not least, I thank my parents Erika and Wolfgang Kramer for their unconditional love and understanding.

Dortmund, January 2008                                        Oliver Kramer

# Contents

## Part II: Self-Adaptive Operators

## Part III: Constraint Handling

## Part IV: Summary



## Part V: Appendix

# 1 Introduction

Evolutionary algorithms (EAs) are biologically inspired, randomized search meta-heuristics. They unify the fundamental principles of biological evolution: inheritance of genes, variation of genes in a population, translation of genotype into phenotype and selection of the fittest in the sense of the Darwinian principle *survival of the fittest* [28]. In the sixties Holland, Rechenberg and Schwefel translated this paradigm of evolution into a concept of algorithms which is called evolutionary computation (EC). Today, this computational method has grown to a rich and frequently used optimization method. It comprises several variants of algorithms which are structurally similar, but specialized to certain search domain characteristics.

## 1.1 Motivation

Today, EA research concentrates on application areas and theoretical questions. A rather unexplained area is the question for parameter settings. EAs exhibit different kinds of parameters, e.g. population sizes or mutation strengths. Their success is bound to the choice of the right parameterizations. Parameters can be *tuned before* the optimization. Human experience is necessary to find the right settings for particular situations. Sometimes it is useful to *control* the parameters *during* the run. A famous example are mutation strengths. At the beginning big mutations accelerate the search. Later on, smaller mutations make the convergence to an optimum possible. Many control techniques are specialized and try to integrate domain knowledge to control the parameters. But flexibility and problem-independence is the spirit of EAs. The question arises: how can EAs *learn* parameters during the optimization automatically? Evolution strategies gave an answer to this question decades ago: self-adaptation. Their mutation strengths self-adaptation became very famous. The parameters are *learned* during the optimization process.

This work concentrates on the famous concept of self-adaptation. What other kinds of parameters can successfully be controlled automatically? How much is evolution able to learn? First, we concentrate on mutation and propose to

O. Kramer: Self-Adaptive Heuristics for Evolutionary Computation, SCI 147, pp. 1–6, 2008.
springerlink.com

control the bias of mutations self-adaptively. Usually, mutations are unbiased, because evolution is not supposed to prefer any direction. But it turned out that a bias is useful, in particular on ridge functions and in constrained domains. Furthermore, we equip inversion mutation with self-adaptation. Inversion mutation changes permutations in combinatorial search domains. In the case of the traveling salesman problem it swaps two edges of random cities. But how many swaps are useful, e.g. at the beginning of the search? We let self-adaptation learn the *optimal* number of swaps in each generation.

Crossover is the other important genetic operator for EAs. It combines the genetic material of two parents. Self-adaptation for crossover has been neglected in the past. We try to investigate whether it is possible to learn automatically which parts of the genetic material to take from each parent. Another interesting question concerns the beginning of self-adaptation: what about the limits of the original mutation strength control of evolution strategies? Experiments and theoretical analysis reveal that their control may fail, in particular at the boundary of infeasible search domain regions. We propose constraint handling heuristics to overcome these limitations.

## 1.2   A Survey of This Book

The following section gives an overview of this book and summarizes each chapter. Figure 1.1 gives a graphical survey of the structure of this work.

*Chapter II: An Introduction to Evolutionary Computation*

The field of EC belongs to the biologically inspired methods of computational intelligence. Evolutionary algorithms model the principle of the famous Darwinian [28] concept of evolution. Various kinds of algorithmic variants exist becoming more and more similar to each other in recent years. Chapter 2 presents the basic principles of the evolutionary computation field, defining basic terms and reaching from practical aspects to a survey of theoretical approaches. Besides a short description of the different types of evolutionary algorithms, the chapter points out the similarities between evolutionary algorithms and particle swarm optimization algorithms.

*Chapter III: An Extended Taxonomy of Parameter Setting Techniques and Definitions of Self-Adaptation*

After the design of the problem representation and the evaluation function, the EA has to be specified. Besides the chose of the genetic operators, adequate parameters for the features of the EA have to be set. The parameter values determine the effectivity and the efficiency of the heuristic. Appropriate operators have to be chosen as well as appropriate initial parameterization of the strategy parameters. Furthermore, it can be useful to change these parameters during the run of the EA. In chapter 3 we introduce and extension of Eiben's [37] parameter control taxonomy. A survey of the various kinds of control techniques

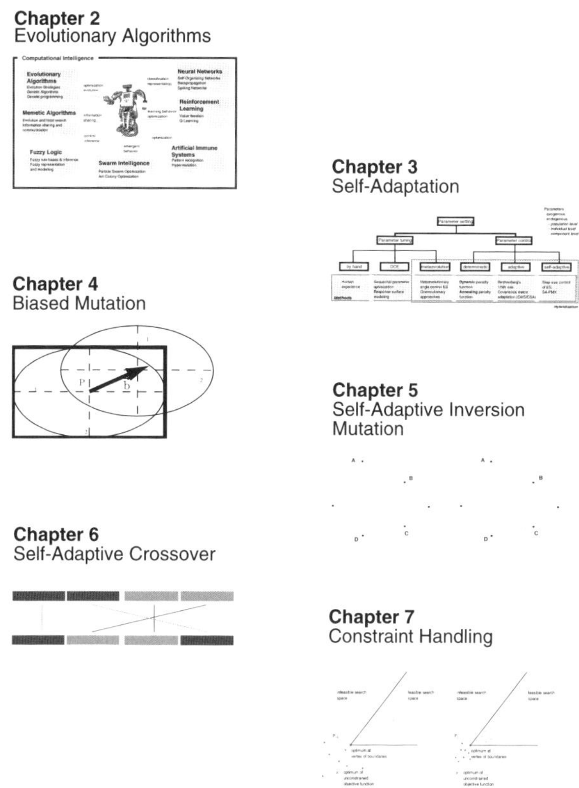

**Fig. 1.1.** Survey of the chapters. After the introductive chapter 2 about EC, chapter 3 describes an extended taxonomy of parameter control and an estimation of distribution view on self-adaptation. In chapter 4 the biased mutation is described and evaluated, while chapter 5 equips inversion mutation with self-adaptation. Chapter 6 answers why the self-adaptation of the crossover structure fails and chapter 7 evaluates methods for optimization at the edge of feasibility.

is complemented by a history of adaptation techniques and an overview over typical adapted parameters. We define self-adaptation in two kinds of ways. In the standard definition self-adaptation is the evolutionary optimization of strategy parameters, which are connected to the objective variables. From the point of view of estimation of distribution algorithms (EDAs) self-adaptation is the evolutionary control of a mixture distribution which estimates the location of the optimum. But self-adaptation also has drawbacks, e.g. the premature step size reduction of ES at the constraint boundary.

*Chapter IV: Self-Adaptive Biased Mutation for Evolution Strategies*

The Gaussian mutation operator of ES exhibits the feature of unbiasedness of its mutations. But offering a bias to particular directions can be a useful degree of freedom, especially when controlled self-adaptively. In order to minimize computational and implementation effort in comparison to other operators, the biased mutation operator (BMO) has been developed. The BMO will be introduced as a self-adaptive operator for biasing mutations into beneficial directions. It makes use of a self-adaptive bias coefficient vector $\xi$ which determines the direction of the bias. The absolute bias can be computed by multiplication with the step sizes. We introduce a couple of variants. The sBMO makes use of only one step size, but offers $N$ directions for the bias while the cBMO uses $N$ steps sizes but only one bias direction. Another biased mutation operator is the DMO, whose bias is adapted according to the descent direction defined by the population centers of two successive generations. Simplified theoretical considerations model the intuition, that the success rate of biased Gaussian mutation is bigger than the success rate of the unbiased correlate on monotone functions. This condition is fulfilled as long as the DMO mechanism is able to find the descent direction. Experiments and statistical tests will show that the bias is beneficial in constrained domains, on ridge functions and on some multimodal problems.

*Chapter V: Self-Adaptive Mutations for Combinatorial Search Domains*

In the past, the most successful applications of self-adaptation concerned mutation operators. The most famous example is the step size adaptation of ES. Discrete strategy parameters for self-adaptive mutations have not been considered much. In this chapter we propose to treat the number of successive inversion mutations as mutation strength and evolve it self-adaptively. Inversion mutation (INV) is a mutation operator for the evolutionary search in combinatorial domains. A famous example is the traveling salesman problem (TSP). One application of INV swaps two pairwise connections, $k$ iterations of INV induce a bigger change of the original solution. The idea of self-adaptive inversion mutation (SA-INV) is the self-adaptive control of the number of iterations the operator INV is applied. A convergence analysis proves that INV and SA-INV find the optimum in finite time with probability one. The experimental results on various TSP instances show that SA-INV accelerates the optimization at the beginning of the search. It turns out that at later phases of the search, more than one applications of INV destroy the solution and lead to significant fitness deteriorations. We call this problem *strategy bound problem* and offer an easy heuristic. The Wilcoxon rank-sum test validates the statistical relevance of the experiments.

*Chapter VI: Self-Adaptive Crossover for Evolutionary Algorithms*

Crossover plays a controversially discussed role within EAs. The building block hypothesis assumes that crossover combines different useful blocks of the solution. The genetic repair effect hypothesis assumes that common features of parental solutions are mixed. The structure of existing crossover operators is usually either fixed or based on randomness. With self-adaptive crossover we take a step towards exploiting the structure of the problem automatically, i.e.

we try to find the building blocks or common features self-adaptively. We introduce various self-adaptive crossover operators for numerical and for string representations. For ES self-adaptive recombination conducts a *morphing* between dominant and intermediate recombination. For genetic algorithms in string representation self-adaptive crossover means the automatic adaptation of crossover points. Our experimental analysis shows the behavior on unimodal, multimodal and constrained search spaces. All experiments will show that neither a significant improvement nor a deterioration can be achieved. We conclude, that the postulated building blocks can hardly be found by means of self-adaptive crossover point control. But in order to show that the optimization of the *structure* of crossover can lead to an improvement, we propose a one-step crossover optimization scheme.

*Chapter VII: Constraint Handling Heuristics for Evolution Strategies*

EAs with a self-adaptive step control mechanism like ES often suffer from premature fitness stagnation on constrained numerical optimization problems. When the optimum lies on the constraint boundary or even in a vertex of the feasible search space, a disadvantageous success probability results in premature step size reduction. A proof of step size reduction leading to premature convergence for a (1+1)-EA under simplified conditions gives insight into success rate situation. We introduce three new constraint-handling methods for ES on constrained continuous search spaces. The death penalty step control evolution strategy (DSES) is based on the controlled reduction of a minimum step size depending on the distance to the infeasible search space. The two sexes evolution strategy (TSES) is inspired by the biological concept of sexual selection and pairing. At last, the nested angle evolution strategy (NAES) is an approach in which the angles of the correlated mutation of the inner ES are adapted by the outer ES. All methods are experimentally evaluated on common test problems and compared with existing penalty-based constraint-handling methods.

*Chapter VIII: Summary and Discussion*

In the last chapter we summarize the main results of this work and discuss its research contributions. Attention is drawn to the benefits and limitations of self-adaptation. Many parameters can be controlled self-adaptively with great success. The mutation step size for ES or the bias of the BMO are examples for successful self-adaptive parameter control. But there are also limitations. Whenever the link between strategy parameter change and fitness is not tight enough, e.g. in the case of self-adaptive crossover, self-adaptation may fail. Strategy parameter drift like premature step size reduction is another negative effect that might occur. At the end, we come to the conclusion that self-adaptation is an advanced feature of EC, which offers great potential, but has to be applied with care and sense for the search space augmentation.

*Experiments*

Many results of this work are based on experiments. Each proposed heuristic or self-adaptive modification is experimentally evaluated. The blocks *experimental settings* show the setup of each experimental series. Our benchmark set of problems, see appendix A, contains a variety of problem types: Typical continuous unimodal functions like sphere and double sum, as well as continuous multimodal functions like griewank and rastrigin have been selected. Hence, the reader can compare our results to own results. We tested most of these problems with $N = 10$ dimensions. For the sake of comparability typical problems in bit string representation, e.g. onemax or ackley, have been chosen. Problems with certain properties like ridge functions or functions with noise in fitness reveal the algorithm's properties on special problem classes. For constrained continuous search domains we have chosen our benchmark functions from the $g$-test suite [83]. The latter is a typical test suite of constrained problems in the EA community. For combinatorial representations we have tested traveling salesman problems of various sizes from the TSPlib [117].

Most of the experiments of this work show the behavior of 25 runs. This, as well as other settings, e.g. initialization interval and termination condition, are oriented to the recommendations of the CEC special sessions on real-parameter optimization 2005 [149] and the session on constrained real-parameter optimization 2006 [83]. Differing termination conditions, i.e. fitness stagnation, have been chosen for problems which suffer from premature step size reduction, see the experiments in chapter 7.

The results are presented in various forms. Tables show the *best, median, worst, mean* and the corresponding standard deviation of all runs, i.e. of the best solution in the last generation. Figures show the development of the best solution in the course of generations. Figures with *typical runs* ought to show *local* properties like cracks or smoothness, which may be lost when averaging. Other figures show the more generalized fitness development on average. With exception of the SA-PMX experiments of chapter 5, *fitness* in the figures means *distance to optimum*. The Wilcoxon rank-sum test proves statistical relevance of many parts of this work. It tests the null hypothesis that the probability distributions of two sample sets are equal. The Wilcoxon rank-sum test is applicable, if the distribution of the data is not known. But it is necessary to point out the following statement.

*All experimental results only show the behavior on the problem classes exemplarily. We are aware that generalization is cautious and due to the no free lunch theorem almost impossible.*

# Part I

---

# Foundations of Evolutionary Computation

# 2 Evolutionary Algorithms

This chapter presents the foundations of the EC framework. After the classification into the methods of computational intelligence and the definition of terms like evolutionary convergence, the basic concepts of EAs are presented. This introduction contains representations, genetic operators and concepts like population models, fitness landscapes and termination conditions. Then the chapter gives an overview of special types of EAs. The following section 2.2 introduces the $(\mu/\rho \overset{+}{,} \lambda)$ - ES, which is the basis of most evolutionary concepts of this work. A survey of theoretical approaches to EC completes this chapter.

## 2.1 Introduction to Evolutionary Computation

EAs are biologically inspired, population based, randomized search metaheuristics. They exhibit the fundamental principles of biological evolution: inheritance of genes, variation of genes in a population, translation of genotype into phenotype and selection of the fittest phenotype in the sense of the Darwinian principle of survival of the fittest. In the sixties and seventies Fogel [44], Holland [58], Rechenberg [114] and Schwefel [133] translated this paradigm of evolution into a concept of algorithms that is called *EC* today. It has grown to a rich and frequently used optimization method. It comprises several variants of algorithms which are structurally very similar, but are specialized to the search domain characteristics.

ES are one of the four main variants of EC techniques. They were invented by Rechenberg [113] and Schwefel [131] in the sixties in Germany and are usually associated with continuous evolutionary optimization. They exhibit operators which are appropriate to real valued search spaces. A more detailed introduction is given later in this chapter. First of all, the chapter starts with an introduction into CI and the EC framework.

### 2.1.1 Computational Intelligence

The main goal of artificial intelligence (AI) is the construction of systems which are able to learn and behave intelligently. Pioneers like John von Neumann

O. Kramer: Self-Adaptive Heuristics for Evolutionary Computation, SCI 147, pp. 9–27, 2008.
springerlink.com © Springer-Verlag Berlin Heidelberg 2008

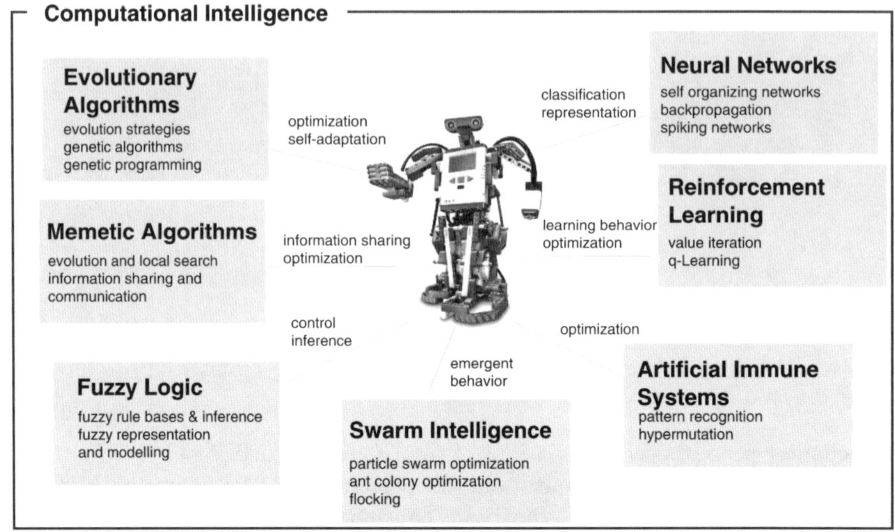


**Fig. 2.1.** Survey of the techniques of CI; photo with permission of LEGO Group

intended to construct more than a single-processor system, but dreamt of intelligent machines even with the possibility of self-reproduction. In the history of AI many approaches were proposed to achieve this objective. Most methods can be classified into two classes: symbolic and subsymbolic approaches. The latter one is also called *computational intelligence* (CI), *soft computing* or *natural computation*. CI comprises neural networks, fuzzy-logic and EAs. Meanwhile, also swarm intelligence, memetic algorithms, artificial immune systems and reinforcement learning can be ranked among CI. These methods also have the feature in common that they are biologically inspired approaches. Figure 2.1 gives an overview of the CI fields.

The field of EC comprises biologically inspired optimization methods and is subject to the work at hand. Memetic algorithms combine evolution with local search. Fuzzy logic enables the representation and modeling of linguistic terms and the inference with rule bases. The field of swarm intelligence comprises a class of algorithms which is inspired by the emergent intelligence of natural swarms. The most successful methods from this class are particle swarm and ant colony optimization. Artificial immune systems are optimization heuristics inspired by processes of the vertebrate immune system like hypermutation and clonal selection. Reinforcement learning algorithms comprise a class of algorithms for the control of agents in dynamical environments. Depending on the available information value iteration or Q-learning are the mostly used variants. Neural networks are inspired by natural neural systems like the human brain. Artificial neural networks are massively parallel systems of neural models, which are able to solve classification problems. Supervised learning methods exist like

Rosenblatt's perceptron or the backpropagation algorithm, which is based on gradient descent in the error function of the classification result. An example for unsupervised neural networks is a self-organizing feature map, which is able to map high dimensional data into low dimensional feature maps. AI and CI methods can be applied to a huge variety of domains, e.g. in the recent past Kramer et al. [76] created a taxonomy of problem classes in the field of AI and music.

### 2.1.2   Optimization Problems and Stochastic Convergence

One of the most important tasks in computer science and AI is optimization. Intuitively speaking, optimization is denoted as as the improvement of parameters of a system. The formal definition is given as follows.

**Definition 2.1 (Optimization Problem)**
*Let $f : \mathcal{X} \to \mathbb{R}$ be the fitness function to be minimized with search set $\mathcal{X}^1$. Find an element $x^* \in \mathcal{X}$ such that $f(x^*) \leq f(x)$ for all $x \in \mathcal{X}$.*

Let $X_t := (x_{t,1}, x_{t,2}, \ldots, x_{t,n})$ be a random population of size $n > 0$ at step $t \geq 0$ and $F_t := \min\{f(x_{t,1}), \ldots, f(x_{t,n})\}$ the best fitness value in the population at step $t$. A desirable property of an optimization method is to find the optimum $x^*$ with fitness $f^*$ within a finite number of steps. This is formalized in the following definition [125].

**Definition 2.2 (Finding an Optimum with Probability One)**
*Let random variable $T := \min\{t \geq 0 : F_t = f^*\}$ denote the first hitting time of the global optimum. An EA is said to visit the global optimum in finite time with probability one if $P(T < \infty) = 1$ regardless of the initialization.*

In some cases an EA does not *find*, but *approximates* the optimal solution or *converges* to $f^*$. As the deterministic concepts of convergence are not appropriate, it is important to introduce the stochastic concepts of convergence as defined by Lukacs [87].

**Definition 2.3 (Stochastic Convergence)**
*Let $D_0, D_1, \ldots$ be non-negative random variables defined on a probability space $(\Omega, \mathcal{A}, P)$. The sequence $(D_t : t \geq 0)$ is said to*

- *converge completely to zero if $\sum_{t=0}^{\infty} P(D_t > \epsilon) < \infty$ for any $\epsilon > 0$,*
- *converge with probability 1 or almost surely to zero if $P(\lim_{t \to \infty} D_t = 0) = 1$,*
- *converge in probability to zero if $P(D_t > \epsilon) = o(1)$ as $t \to \infty$ for any $\epsilon > 0$,*
- *and to converge in mean to zero if $E[D_t] = o(1)$ as $t \to \infty$.*

With this concept, Rudolph [125] defines the term *convergence* for EAs.

---

[1] The problems we consider in this work are minimization problems. If we speak of *high* fitness, we mean *low* values.

**Definition 2.4 (Evolutionary Convergence)**
*Let $(X_t : t \geq 0)$ be the sequence of populations generated by some EA and let $F_t = \min\{f(X_{t,1}), \ldots, f(X_{t,n})\}$ denote the best objective function value of the population of size $n < \infty$ at generation $t \geq 0$. An EA is said to converge completely (with probability 1, in probability, in mean) to the global minimum $f^* = \min\{f(x); x \in \mathcal{X}\}$ of objective function $f : \mathcal{X} \to \mathbb{R}$ if the nonnegative random sequence $(D_t : t \geq 0)$ with $D_t = f^* - F_t$ converges completely (with probability 1, in probability, in mean) to zero.*

Rudolph comments that a precondition for convergence is the property of an EA to visit the global solution $f^*$. But the property of convergence does not indicate an advantage of finding the global optimal solution automatically. Furthermore, we need the concept of success rate or success probability. This denotes the rate of improved solutions and all solutions produced by a genetic operator $OP$.

**Definition 2.5 (Success Rate, Success Probability)**
*Let $x_t$ be an individual at generation $t$ and $\mathcal{A}_t = \{y | y = OP(x_t)\}$ the set of samples produced by the genetic operator $OP$. Let $\mathcal{S}_t = \{z \in \mathcal{A}_t | f(z) < f(x_t)\}$, $\mathcal{S}_t \subseteq \mathcal{A}_t$ be the set of samples with better fitness than $x_t$. The success rate or success probability $(p_s)_t$ is defined as ratio of the successful offspring individuals $\mathcal{S}_t$ and all offspring individuals $\mathcal{A}_t$*

$$(p_s)_t = \frac{|\mathcal{S}_t|}{|\mathcal{A}_t|}. \tag{2.1}$$

These definitions will be useful for theoretical considerations, in particular in chapter 4, 5 and 7.


### 2.1.3   Classic Optimization Methods

The classical non-evolutionary optimization methods for continuous problems can mainly be classified into *direct, gradient* and *Hessian* search methods. The direct methods determine the search direction without using a derivative [135], similar to EAs. Lewis et al. [82] give an overview of direct search methods. Pattern search methods [29] examine the objective function with a pattern of points which lie on a rational lattice. Simplex search [99] is based on the idea that a gradient can be estimated with a set of N+1 points, i.e. a simplex. Direct search methods like Rosenbrock's [121] and Powell's [110] collect information about the curvature of the objective function during the course of the search. If the derivatives of a function are available, the gradient and Hessian methods can be applied. Gradient methods take the first derivative of the function into account, while the Hessian methods also compute the second derivative. Successful examples are the quasi-Newton methods [21]. They search for the stationary point of a function, where the gradient is 0. Quasi-Newton estimates the Hessian matrix analyzing successive gradient vectors. In practice, the first derivative is often not available. This and other requirements of many classic search methods are the reasons for the success of EAs.

### 2.1.4   A Short Excursus to Molecular Biology and Genetics

Carrier of the genetic information of organisms is the DNA, deoxyribonucleic acid [156]. Nucleic acids are linear unbranched polymers, i.e. chain molecules, of nucleotides. Nucleotides are divided into purines (adenine - A, guanine - G) and pyrimidines (thymine - T, cytosine - C). The DNA is organized into a double helix structure. Complementary nucleotides (bases) are pitted against each other: A and T, as well as G and C. The next level of structure are the codons. A sequence of three nucleotides is a codon, also called triplet. Because a codon consists of three bases, $4^3 = 64$ different codons exist. With three exceptions all codons code one of 20 amino acids. The synonyms code identical amino acids. The DNA macro molecules are wrapped around large protein molecules, the histones. Such a structure is called nucleosome, the histone molecule is the nucleosome core. The chain of nucleosomes is called chromatin. The rewind of the DNA results in a compression of 1:7. A further compression into chromatin chains achieves a space reduction by $10^{-4}$. The human genome is about 3 billion base pairs long, arranged in 23 pairs of homologous chromosomes. All base pairs would have an overall length of 2.6 m, but are compressed in the core to size of $200\mu m$.

Now, what is a gene? In fact, the definition has not satisfactorily been given until now. The most popular definition is: A gene as a DNA sequence that codes for a particular protein. But several thousands of genes have been found that do not code for proteins, but seem to have other tasks. A recent estimation of the number of human genes is 20488 [107]. The transformation from genotype to phenotype is called gene expression. In the first phase, the transcription, the DNA is translated into the RNA. In the second phase, the translation, the RNA synthesizes proteins. Proteins are polypeptide chains consisting of 20 types of amino acids. Amino acids consist of a carboxyl and an amino group and differ in their rest group that may also be aromatic, i.e. contain the famous hexagonal benzene molecule. The peptide bound of the long polypeptide chains happens between the amino and the carboxyl group of the neighbored molecule. Proteins are the basis modules of all cells. They build characteristic three dimensional structures, e.g. the alpha helix molecule.

### 2.1.5   Concepts of the Evolutionary Computation Framework

The essential idea of EC is the randomized exploration of the search space with a population of candidate solutions, which are improved during each iteration of the algorithm. These iterations are called generations in terms of EC. The improvement of the candidate solutions is achieved in two steps. In the first step, the existing solutions are diversified with random variation. In the second step, the best solutions are selected and provide a direction for the search. There exist various kinds of genetic operators which are designed for the various types of representations, as well as search space features. In chapters 4 and 6 the most important types of genetic variation operators, mutation and crossover, are presented. The EC methods are supposed to find solutions which satisfy certain quality criteria.

**Exploration and Exploitation**

All EC methods have the following two basic principles in common:

- *Exploration* of the search domain in order to find the optimum. The exploration part of an EC algorithm is covered by the genetic operators mutation and crossover. Mutation is supposed to diversify existing candidate solutions with means of random variation. Crossover or recombination on the one hand produces new candidate solutions by combining features of at least two solutions. On the other hand, it hereby *exploits* discovered knowledge, which is the essential part of the second key feature:
- *Exploitation* of knowledge about the fitness landscape which could have been discovered within the past optimization process. The selection operators achieve this objective by choosing the fittest mates for reproduction (parent selection) or by letting the fittest solutions survive (survival selection).

**Objective and Strategy Variables**

In the language of EC candidate solutions are called individuals. The solution is called phenotype, while the representation of a candidate solution is called genotype. It consists of objective variables on which the optimization heuristic works. Canonical genetic algorithms work on binary representations of objective variables, i.e. $\mathbf{b} = (b_1, \ldots, b_N) \in \{0, 1\}^N$, where $b_i$ denotes a gene and $N$ is the length of the chromosome. A Population $\mathcal{P}$ is a set of chromosomes. For a gene $b_i \in \{0, 1\}$, the index i denotes the locus of $b_i$, whereas the potential values 0 and 1 are alleles of $b_i$. Numerical or continuous representations, e.g. ES, work with vectors of continuous objective variables $\mathbf{x} = (x_1, x_2, \ldots, x_n) \in \mathbb{R}^N$. Representations for combinatorial search spaces are versatile. The most famous ones are simple ordering problems, where the quality of the solution only depends on the order of the solution's components. A typical combinatorial problem is the traveling salesman problem (TSP) [45], which is also used as a test problem in chapter 6. Some problems might require a mixture of the mentioned representations.

**Mutation**

Mutation is the source for evolutionary changes. In biology mutation of the DNA occurs during copy processes or due to influences of the environment. Offspring individuals are not exact copies of their parents, but the variation is also a source for innovation. In binary representations mutation is bit flipping with a certain probability. For numerical search spaces mutation can be implemented as an addition of a random number based on a certain probability density function. In genetic programming (GP) mutation is the exchange of a subtree by a random subtree. Mutation produces small changes in the chromosome with high probability, bigger changes with decreasing probability. Mutation operators ought to fulfill the three established rules of reachability, unbiasedness and scalability [16], see also chapter 4 where in particular the second rule is discussed. The strength of mutation is also known as step size or mutation rate and is a measure for the level of change of the solutions.

## Crossover/Recombination

The role of crossover, also called recombination, is discussed intensively. Recombination aims at the inheritance of genetic material from at least two parental solutions. In chapter 6, the role of recombination is discussed. For string representations, a crossover point is chosen (*one-point crossover*) and two offspring solutions are created by exchanging the two parts of the divided strings. For numerical representations, crossover can be implemented as the arithmetic median of the continuous objective variables. The *building block hypothesis* (BBH) [47] claims that recombination contributes to the identification of useful solution constructing blocks. In contrast to the BBH, the theory of the *genetic repair effect* claims that recombination identifies and combines common features of the solution.

## Selection

Selection is the antagonist to the variation operators. It gives the evolutionary search a direction. Only selected individuals are allowed to survive or inherit their genetic material to the offspring. There are two types of selection operators of which the first one mentioned here is *parent selection*. In order to produce offspring via recombination, a number of parents have to be selected. Parent selection is equivalent to mate selection in biology. Its implementations can reach from random selection to fitness based selection schemes. A typical selection mechanism is called *fitness proportional selection*, where the probability $p_a$ that an individual $a$ is selected depends on its absolute fitness compared to the absolute fitness of the rest of the population

$$p_a = \frac{f_a}{\sum_j f_j} = \frac{f_a}{\overline{f}n} \tag{2.2}$$

with the average population fitness $\overline{f} = \frac{1}{n}\sum_j f_j$ For the case of standard GAs, where the population size is fixed, the expected number of copies of each individual becomes

$$n_a = p_a n = f_a/\overline{f}. \tag{2.3}$$

The second type of selection operators is *survivor selection*. After the offspring population has been generated, survivor selection gives a direction to the optimization process by copying the selected individuals into the population of the next generation. This concept is equivalent to Darwin's principle of survival of the fittest.

**Fitness**

The quality of a candidate solution is called fitness referring to the biological concept of fitness. The fitness of a solution can be derived by

- *real world features* - the quality of a real-world system can be measured by a set of defined features, which have to be translated into comparable values,
- *simulation models* - here, systems are simulated artificially and analog to online optimization,
- *mathematical models*, and
- *test suite problems* for experimental EC studies.

**Algorithm Structure and Populations Models**

Figure 2.2 shows the pseudocode of a generalized EA. In the first step a population of so called individuals which represent a solution to an optimization or search problem is initialized and evaluated. In an iteration loop the population of solutions is changed, evaluated and selected until a termination condition is fulfilled. Meanwhile, there exist various classes of EAs, partly with different historical background. They differ in the kind of representation, genetic operators and problem domains they are appropriate to.

The two main variants of *panmictic* population models are the generational model and the steady-state model. In the generational model $\lambda = \mu$ individuals are produced from the parental population of size $\mu$ using the genetic operators. Afterwards, all offspring individuals replace the parents of the last generation. In the steady-state model not the entire population is changed, but $\lambda < \mu$ new individuals replace $\lambda$ old ones. The ratio $\lambda/\mu$ is called generational gap. All individuals in such a panmictic population model are considered to be potential mates. Recently, many other population models were introduced, which are based on neighborhood relationships. In such structured population models individuals can only interact with certain neighbors determined by these relations. In the *island* or *multi population models* the panmictic population is divided into several smaller ones [154].

| | |
|---|---|
| 1 | **Start** |
| 2 | t:=0; |
| 3 | Initialize $\mathcal{P}(t)$; |
| 4 | Assign fitness to $\mathcal{P}(t)$; |
| 5 | **Repeat** |
| 6 | Variation of $\mathcal{P}(t) \rightarrow \mathcal{P}'(t)$ |
| 7 | Assign fitness to $\mathcal{P}'(t)$; |
| 8 | Select $\mathcal{P}(t+1)$ from $\mathcal{P}'(t)$; |
| 9 | $t := t + 1$; |
| 10 | **Until** termination condition |
| 11 | **End** |

**Fig. 2.2.** Pseudocode of a generalized evolutionary algorithm

**Termination Conditions**

When an EA should terminate depends on the demands defined by the user and on practical conditions. Typical termination conditions for EAs are

- the fitness has reached a predefined quality,
- fitness convergence, i.e. no fitness improvement is achieved after a predefined number of generations,
- objective variable convergence, which is usually equivalent to fitness convergence in non-dynamic search domains
- strategy variable convergence,
- maximum number of generations or fitness function calls is reached, or
- maximum amount of time has passed.

The termination condition must be defined accurately for the comparison of approaches.

**Fitness Landscapes**

Several kinds of fitness landscapes with certain features exist. In the following, the most important types of fitness landscapes are introduced.

- *Unimodal* - an unimodal fitness landscape owns only one global optimum.
- *Multimodal* - in contrast to unimodal fitness landscapes, multimodal own multiple local optima and one or more global optima. The danger for EC algorithms is to get stuck in local optima without being able to find the global one.
- *Dynamic* - in dynamic fitness landscapes, the fitness changes over time. So, the global optimum may be moving. The task of an optimizer is to find and follow the moving optimum.
- *Constrained* - in constrained fitness landscapes, parts of the search space are infeasible. Several problems for an EC algorithm occur. In heavily constrained search spaces feasible solutions have to be created. Feasible parts of the search space may not be connected and it can be difficult for an EA to jump between these parts. Another problem is the situation when optima lie at the boundary of the infeasible search space, in particular when they are located in a vertex of the feasible space. This problem is described and analyzed in chapter 7.
- *Multi-objective* - for multi-objective optimization problems not only one, but a whole set of objectives has to be optimized. One task may be to identify a Pareto set of non-dominated solutions.

For each fitness landscape heuristic modifications of EAs exist.

## 2.1.6 Types of Evolutionary Algorithms

The following part gives a brief outline of common types of EAs. As already mentioned, they grow together. The subsequent section will introduce ES.

## Evolutionary Programming

Evolutionary programming (EP) is similar to evolution strategies (ES) with respect to certain features. In its original form Fogel, Owens and Walsh [44] developed EP for higher levels of abstraction. They were designed to produce deterministic finite automatons, intended to match input to corresponding output data. A typical EP algorithm for real-valued representations works with vectors and Gaussian perturbation as means of mutation, but without recombination. The so called *meta-EP* makes use of self-adaptive mutation which works as follows: The chromosome $\mathbf{a} = (x_1, \ldots, x_N, \sigma_1, \ldots, \sigma_N)$ is mutated into the successor $\mathbf{a}'$ using the following equations

$$x'_i = x_i + \sigma'_i \cdot N_i(0, 1), \tag{2.4}$$

$$\sigma'_i = \sigma_i \cdot (1 + \alpha \cdot N(0, 1)). \tag{2.5}$$

Usually, the parent selection takes place deterministically while the survivor selection is a probabilistic.

## Genetic Algorithms

Genetic algorithms (GAs) were initially invented by John Holland in the USA in the sixties for adaptation in natural and artificial systems [58]. Today, the term *genetic algorithms* is most frequently used for EC algorithms when thinking of a string representation of the individuals. The so called canonical or simple GA works as follows. From a population $\mathcal{P}$ of $\mu$ individuals $\mu$ new individuals are produced by selecting two parents from $\mathcal{P}$ with fitness proportional selection. In a second variation step, these $\mu$ individuals are mutated, e.g. with uniform mutation and afterwards put into the population $\mathcal{P}'$. There exists a variety of genetic operators for the different representation types. Some of them are introduced in later chapters of this book. We evolved a reactive rule base for the control of a human-competitive multi player computer game using GAs [111], [112], see section 2.3.

## Genetic Programming

Genetic programming (GP) is the youngest branch of EC methods. GP is about the automatic evolvement of computer programs, e.g. to control agents or robots. Although this idea was not considered to be new, John Koza is seen as its inventor [70]. The idea is to evolve programs by means of the genetic operators mutation, crossover and selection of the qualitatively most successful models. The most famous representation of computer programs for GP are trees, but programs can also be represented in machine language like assembler. The parameter sets of a GP program are called Koza-Tableau. Mutation in tree representation means the replacement of an existing subtree by a randomly generated subtree, which can be bound in depth. Crossover exhibits a greater importance than mutation in GP. In most cases it is run with probability 0.9. Crossover means the exchange of subtrees of two parents. GP programs may result in big trees. In order to avoid the generation of frequently used subprograms, the so called automatically defined functions (ADFs) were introduced, which are frequently evolved subprograms. An effect often observed in GP is bloat [141]. It means that big programs are result of the evolutionary process. This effect is also called *survival of the fattest*.

## Particle Swarm Optimization

Why do particle swarm optimization algorithms (PSOAs) and EAs both belong into the field of EC? They follow similar concepts of exploration and exploitation of the search space. In particular, the continuous PSOA is very similar to a $(1+\lambda)$-ES, see section 2.2. The algorithmic structure as well as the operators are quite similar. As a consequence, concepts from EAs can extend particle swarm optimization, e.g. self-adaptation. Furthermore, genetic operators inspired by particle swarm optimization can be designed.

Particle swarm optimization is a population based optimization approach, which was introduced by Kennedy and Eberhart [69]. It was originally introduced as an alternative to the standard GA. Particle swarm optimization was inspired by insect swarms, as the latter consist of a number of particles moving in the search space. Similar to ES each particle $\mathbf{x}$ represents a candidate solution for a numerical problem. Each particle is assigned with a velocity $\mathbf{v}$. The following two equations exhibit the working principle of PSOAs:

$$\mathbf{x}' = \mathbf{x} + \mathbf{v}', \tag{2.6}$$

and

$$\mathbf{v}' = \mathbf{v} + \phi_1(\mathbf{pb} - \mathbf{x}) + \phi_2(\mathbf{b} - \mathbf{x}) \tag{2.7}$$

The variables $\phi_1$ and $\phi_2$ produce random numbers, whereas $\mathbf{pb}$ is the best solution of the particle's history and $\mathbf{b}$ is the best solution of all particles' histories. Several enhancements of the above PSOA equations have been proposed. E.g. the velocity can be extended by an inertia term as proposed by Shi and Eberhart [139]

$$\mathbf{v}' = w\mathbf{v} + \phi_1(\mathbf{pb} - \mathbf{x}) + \phi_2(\mathbf{b} - \mathbf{x}). \tag{2.8}$$

Shi and Eberhart propose to decrease this term if no improvement is obtained within consecutive time steps. They also proposed to replace the worst particle by an elite particle, the best of the whole swarm. There have been attempts to evolve particle swarm equations with genetic programming [108].

## Similarities Between EAs and PSOAs

Like EAs, PSOAs are population based heuristics that contain randomized aspects. Table 2.1 summarizes the comparison between EA and PSOA concepts.

**Table 2.1.** Comparison of typical EA and PSOA concepts

| concept | EA | PSOA |
| --- | --- | --- |
| candidate solution | individual | particle |
| set of solutions | population | swarm |
| variation (exploration) | mutation | randomized factors $\phi_1, \phi_2$ |
| variation (exploitation) | crossover | equation parts $(\mathbf{pb} - \mathbf{x})$, $(\mathbf{b} - \mathbf{x})$ |
| direction of search | parent & survivor selection | best individuals $\mathbf{pb}$, $\mathbf{b}$ |

Both work with a set of candidate solutions. In terms of EAs they are called population of individuals, in terms of PSOAs they are denoted as a swarm of particles. This similarity is not restricted to the algorithmic scheme, but furthermore holds for the used *operators*. The three genetic operators mutation, crossover and selection can be identified within the PSOA equations. We compare them to the equations of the $(1+\lambda)-\mathrm{ES}$. In a $(1+\lambda)$-ES the best individual $\mathbf{b}$ is chosen as the only parent for creating $\lambda$ offspring solutions. The equation of a $(1+\lambda)$-ES looks as follows

$$\mathbf{x}' = \mathbf{b} + \phi \tag{2.9}$$

with mutation $\phi$. It is structurally very similar to the PSOA equation 2.10. Assuming that $\mathbf{pb}$ and $\mathbf{b}$ are almost identical, the PSOA equation can be written as

$$\mathbf{x}' = (\phi_1 + \phi_2)\mathbf{b} + (1 - \phi_1 - \phi_2)\mathbf{x} + \mathbf{v}. \tag{2.10}$$

Hence, the new solution is built by combining the old solution with the best one - or at least one of the best solutions - weighting the ratios with randomized parameters.

*Mutation.* Mutation, i.e. exploration of the search space, is achieved by multiplication with the random numbers $\phi_1$ and $\phi_2$. In ES mutation of objective variables is usually achieved by adding a random value based on Gaussian mutation. But multiplicative mutation is also known for strategy parameters, e.g. for correlated mutation [18] or biased mutation, see section 4.2.

*Crossover.* Crossover or recombination can be found in equation 2.10 as the new solution is the weighted sum of the best particle (or the best particle and the best position of one particle's history) and its last position

$$\mathbf{x}' = \alpha_1 \mathbf{b} + \alpha_2 \mathbf{x} + \mathbf{v} \tag{2.11}$$

with $\alpha_1, \alpha_2 \in \mathbb{R}$. Hence, a PSOA is an EA in which every individual of the population survives and is recombined with the best individual of the population.

*Selection.* In the equation 2.7 selection takes place: in the PSOA the best solution found so far is used for the construction of every new candidate solution. This is similar to the survival of the best individual of the elitist plus-selection scheme. But the elitist selection scheme is softened as not only the best solution **b** found so far is used, but also to some extend the best solution **pb** of a particle's history. The comparison only reveals some structural similarities and relationships between PSOAs and ES. But it is neither a prove for identical convergence properties nor allows reasoning concerning the behavior on certain function classes.

## 2.2  Evolution Strategies

ES are one of the four main variants of EAs besides GAs, EP and GP. They were invented by Rechenberg and Schwefel in the middle of the sixties at the Technical University of Berlin [114], [134]. ES can successfully be applied to engineering problem domains. In this section the basic features of the so called $(\mu/\rho \overset{+}{,} \lambda)$-ES are introduced.

### 2.2.1  The $(\mu/\rho \overset{+}{,} \lambda)$-ES

For a comprehensive introduction to ES see Beyer and Schwefel [18]. Here, the most important features of the state of the art $(\mu/\rho \overset{+}{,} \lambda)$-ES[2] for continuous search spaces are repeated. An ES uses a parent population with cardinality $\mu$ and an offspring population with cardinality $\lambda$. Each individual consists of objective and strategy variables $\mathbf{a} = (x_1, \ldots, x_N, \sigma_1, \ldots, \sigma_N, F(x))$, with problem dimension $N$. At first, the individuals are initialized. The objective variables represent a potential solution to the problem whereas the strategy variables, which are step sizes in the standard $(\mu/\rho \overset{+}{,} \lambda)$-ES, provide the variation operators with information how to produce new results. During each generation $\lambda$ individuals are produced in the following way. In the first step $\rho$ $(1 \leq \rho \leq \mu)$ parents are randomly selected for reproduction. After recombination of strategy and objective variables, the ES applies log-normal mutation to the step sizes $\boldsymbol{\sigma} = (\sigma_1, \ldots, \sigma_N)$ and uncorrelated Gaussian mutation to the objective variables, see also section 3.4.

$$\mathbf{x}' := \mathbf{x} + \mathbf{z} \tag{2.12}$$

$$\mathbf{z} := (\sigma_1 \mathcal{N}_1(0,1), \ldots, \sigma_N \mathcal{N}_N(0,1)) \tag{2.13}$$

$$\boldsymbol{\sigma}' := e^{(\tau_0 \mathcal{N}_0(0,1))} \cdot \left(\sigma_1 e^{(\tau_1 \mathcal{N}_1(0,1))}, \ldots, \sigma_N e^{(\tau_1 \mathcal{N}_N(0,1))}\right) \tag{2.14}$$

---

[2] The $\overset{+}{,}$-notation combines the notation for the selection schemes plus and comma selection.

| | |
|---|---|
| 1 | **Start** |
| 2 | t:=0; |
| 3 | Initialize $\mathcal{P}(t)$; |
| 4 | **Repeat** |
| 5 | **For** k=1 **To** $\lambda$ **Do** |
| 6 | Choose $\rho$ parents from $\mathcal{P}(t)$ |
| 7 | $\sigma_k$:=recombination_strategy_variables; |
| 8 | $\mathbf{x}_k$:=recombination_objective_variables; |
| 9 | $\sigma_k$:=mutation_strategy_variables; |
| 10 | $\mathbf{x}_k$:=mutation_objective_variables; |
| 11 | $F_k := f(\mathbf{x}_k)$; |
| 12 | **Next** |
| 13 | $\mathcal{O}(t) := \{(\mathbf{x}_1, \sigma_1, F_1), \ldots, (\mathbf{x}_\lambda, \sigma_\lambda, F_\lambda)\}$; |
| 14 | Select $\mathcal{P}(t+1)$ from $\mathcal{O}(t) \cup \mathcal{P}(t)$ in the case of $(\mu + \lambda)$-selection, |
| 15 | or from $\mathcal{O}(t)$ in the case of $(\mu, \lambda)$-selection; |
| 16 | $t := t + 1$; |
| 17 | **Until** termination condition |
| 18 | **End** |

**Fig. 2.3.** Pseudo code of a $(\mu/\rho \overset{+}{,} \lambda)$-ES

After $\lambda$ individuals are produced, the best $\mu$ individuals are selected as parents for the next generation exclusively from the offspring population in case of *comma selection* or from the offspring and the previous parental population in case of the *plus selection* scheme. Figure 2.3 shows the pseudocode of a $(\mu/\rho \overset{+}{,} \lambda)$-ES.

### 2.2.2 The $(\mu, \kappa, \lambda, \rho)$-ES

The comma selection limits the maximal lifespan of an individual to one generation, while an individual has an unlimited lifespan under plus selection. This is not necessarily the case. The parameter $\kappa > 0$ introduces a maximal number of generations an individual is allowed to survive. After $\kappa$ generations the individual will be replaced, even if no better or equal solution exists. This variant of ES is called $(\mu, \kappa, \lambda, \rho)$-ES. The TSES in chapter 7 makes use of the $\kappa$-parameter.

## 2.3 Practical Guidelines for Evolutionary Algorithms

In the following we summarize some practical EA wisdoms, which haven proven their worth in recent projects.

- The *representation* should be chosen wisely as on the one hand it determines the size of the search space. On the other hand, the representation has a direct influence on the impact of genetic operators.
- The *population sizes* control the *diversity*. Small population sizes might be efficient, but often have to be paid with a decrease in quality of the results (efficiency vs. accuracy).

- The *selection pressure* is strongly connected to the population sizes. A high selection pressure can increase the optimization speed, but otherwise increases the probability of getting stuck in local optima. Again, we have the tradeoff between efficiency and accuracy.
- The *mutation strength* influences the ability of the algorithm to converge: a high mutation strength is advantageous at the beginning of the search. But when approximating the optimum, the mutation strength should decrease, so that the EA can converge. If the mutation strength is too small at the beginning, the search will be slow.
- The useful scope of the mutation strength is not easy to guess. A *self-adaptive control* of the mutation strength frees the mutation from under- or oversized steps, see chapter 3.
- According to the genetic repair effect, *crossover* is useful to mix the *common* features of parents, see chapter 6. Hence, a crossover operator should be designed to contribute to this demand.
- Concerning the *initialization*, the practitioner should ask himself, how much information is available at the beginning of the search? If the first individuals can be initialized with close-to-optimum solutions, the search is only a small optimization process of diminutive improvements. For this case, only a small mutation strength is recommendable not to destroy close to optimum solutions. A *from the scratch* search requires a sophisticated representation and choice of the operators and their parameterization.

This is only a short and incomplete survey of practical guidelines for the usage of EAs. The section 2.4 gives an overview of theoretical EA research fields.

**Example of an Application: Evolution of a Computer Game Opponent**

We tested the capabilities of EAs for building a rule base for a dynamic computer game opponent, based on imitation [111] and learning from scratch [112]. The basis of our experiments was a simple scenario. Two artificial computer opponents in a simple first person shooter scenario *fight* each other for approximately one minute. The EA driven opponent selects his rules from a rule list $(R_1, \ldots, R_s)$ of fixed size $s$. We tested several settings, reaching from $s = 10$ to $s = 400$. Each rule $R_i$ consists of a matrix of bits, representing the environment divided into grids and a corresponding action. By means of the genetic operators the population of rule bases was evolved and evaluated letting the artificial opponent play for about one minute. In each step, the current environment is compared to the rule base and the rule with the least Euclidean distance to the current situation is executed. The fitness of the whole rule base is calculated by simply *summing up* the hits $h_{\mathrm{EA}}$, the EA agent achieves during the play, *subtracting* the hits $h_\mathrm{o}$ his opponent achieves.

$$f_t = w_{\mathrm{EA}}(h_{\mathrm{EA}})_t + w_\mathrm{o}(h_\mathrm{o})_t \tag{2.15}$$

The weights $w_{\mathrm{EA}}$ and $w_\mathrm{o}$ allow to control the aggressiveness of the artificial player. In a first approach rules were initialized keeping track of human or artificial players [111]. Later we evolved the rules based on random initializations [112]. In both approaches mutation was implemented as bit flipping of the grid entries. Recombination combines the rules of two different rule bases of the population. The experiments showed that the evolved players were in most cases comparable or better than the original artificial players of the computer game. We successfully experimented with coevolution, i.e. the training of two EA based players at the same time.

## 2.4  Theories of Evolutionary Algorithms

The theoretical analysis of evolutionary techniques is suffering from the huge complexity of their many random factors. Current research tries to investigate at least little insight into predictive models for the behavior of EAs as well as providing the means for the design of efficient optimizers for given problem classes. In the EA community there is still a lot of discussion about the relation of theoretical and experimental analysis. The work at hand is not restricted to the experimental analysis of the proposed methods. This is why we include this survey of theoretical EA research. Theoretical research about EAs can be divided into the following main parts: no free lunch, building block hypothesis, dynamical systems analysis, statistical systems analysis, Markov chains and runtime analysis.

### No Free Lunch

In the no free lunch (NFL) theorem Wolpert and Macready [161] showed that, if we average over the space of all possible problems, all algorithms have the same performance. This statement is feasible for non-revisiting algorithms with no problem-specific knowledge. One of the important questions is whether the set of practical problems is representative for all problems or whether it is a certain subset for which the NFL theorem does not hold. At least algorithm engineers should be aware that *award-winning* algorithms on some problems will show poor results on other problems.

### Holland's Schema Theorem:

A schema is a set of individuals with undefined genes at defined locations. It can be seen as a hyperplane in the search space. Using the "don't care" symbol #, 10#### is an example for a binary schema, 100101 is one possible instance of this schema of $2^4$ possible ones in this case. The schema-theorem states that schemata with high fitness increase their number of instances within the population from generation to generation [38]. Let $m(H, z)$ be the proportion of individuals representing schema $H$ at time $z$. The schema theorem states that $m(H, t + 1) > m(H, t)$, i.e.

$$m(H, t + 1) \geq m(H, t) \cdot \frac{f(H)}{<f>} \cdot \left[1 - \left(p_c \cdot \frac{d(H)}{l - 1}\right)\right] \cdot [1 - p_m \cdot o(H)] \quad (2.16)$$

with $f(H)$ representing the fitness of schema $H$, the mean population fitness $< f >$, the defining length $d(H)^3$ and the order $o(H)^4$, the probability $p_c$ of applying crossover and the probability $p_m$ to apply bitwise mutation.

### Building Block Hypothesis

The analysis of the schema theorem has led to the BBH [47]. Broadly speaking the BBH assumes that low-order schemata compete among each other and combine to higher order schemata until the globally optimal solution is produced. Goldberg claims that the BBH is supported by the above mentioned schema theorem. Experiments have been performed which seem to be inconsistent with the BBH, i.e. they show that uniform crossover performs better than 1-or 2-point crossover. But uniform crossover has a very disruptive effect on strings and blocks. In chapter 6 we try to *catch* the building blocks with self-adaptive crossover.

### Dynamical Systems Analysis

In order to describe the amelioration success of an ES local performance measures are necessary. They are expected values of population states and enable the evaluation of the local amelioration power of the ES, i.e. the success between two consecutive generations [18]. The *quality gain* measures the progress in the fitness space whereas the *progress rate* is a local performance measure in the object parameter space. The *progress rate* $\varphi$ is defined as the expected distance change $\Delta r$ of the parental population centroid toward the optimum $\hat{\mathbf{y}}$ in the object parameter space [18]

$$\varphi := E[\Delta r] = R^{(g)} - R^{(g+1)} \tag{2.17}$$

with

$$\Delta r := r^{(g)} - r^{(g+1)} \tag{2.18}$$

and

$$r^{(.)} := \left\| \langle \mathbf{y} \rangle_p^{(.)} - \hat{\mathbf{y}} \right\|. \tag{2.19}$$

The capital $R = E[r]$ denotes the expected value of the lower case letter $r$ denoting a random variable. Calculating $\varphi$ is difficult and therefore mainly analyzed for the sphere function $F(\mathbf{y}) = c \|\mathbf{y}\|^{\alpha}$, $\mathbf{y} \in R^N$ and for the ridge function class [15]. The sphere function is totally symmetric and its fitness values only depend on the distance $R$ to the optimum. Hence, the one-dimensional dynamic of $R$ reduces dynamics of $N$ dimensions. Using an ES with isotropic Gaussian

---

$^3$ The defining length $d(H)$ of a schema is the distance between the outermost defined positions of the schema parts.

$^4$ The order $o(H)$ of a schema is the number of positions which do not exhibit the # symbol.

mutations $\sigma$ and a given parental state $R$ [18], the progress rate $\varphi$ only depends on the parameters $R$, $N$ and $\sigma$ in the following way:

$$\varphi^* := \varphi\frac{N}{R} \text{ and } \sigma^* := \sigma\frac{N}{R} \tag{2.20}$$

Using these normalizations it can be shown that the normalized progress rate $\varphi^*$ depends only on the normalized step size $\sigma^*$ and the search space dimension $N$. But as the dependency on $N$ is quite weak the asymptotic case $N \to \infty, \varphi^* = f(\sigma^*)$ often suffices [18].

Michael Vose's dynamical systems approach belongs to the infinite population models [155]. This approach models EAs in infinite search spaces. It makes use of a vector whose components represent the proportion of the population of a certain type. A mixing matrix represents the operators mutation and recombination while a selection matrix represents the selection operator and the objective function. Research concentrates on the formulation of models for the matrices and various interesting results can be gathered from this kind of analysis.

### Statistical Mechanics Analysis

The statistical mechanics analysis is a macroscopic approach assuming that a number of variables influence the whole system similar to the modeling of complex systems in physics [138]. It allows the analysis of the mean and variance of a population. Although its results are very accurate predicting the behavior of real EAs the analysis is restricted to the above mentioned features.

### Markov Chain Analysis

The evolutionary system can be seen as a stochastic process [38], i.e. as a time-discrete Markov chain if a couple of conditions are fulfilled. Each state represents the different possible populations. Among the conditions is that the system can at any time be characterized by a certain state and that the state transitions depend on the current state exclusively regardless from the past sequence of states. From this condition a transition matrix $M$ can be formulated whose entry $M_{ij}$ gives the probability to move from state $i$ to state $j$ in one step. The $(i,j)$-th entry of the matrix $M^n$ contains the probability to move from state $i$ to state $j$ in $n$ steps. Markov chains have been analyzed extensively in the past and many results from this analysis can be applied to EAs. An example is the result of Rudolph [124]: a GA with nonzero mutation and elitism will always converge to the global optimum, but not necessarily without elitism.

### Runtime Analysis

The runtime analysis of EAs concentrates on the analysis of the number of generations until the optimum can be found. The research concentrates on showing upper and lower bounds for the expected runtime of an EA and the success probability. Various proof techniques have been developed to bound the runtime of EAs, e.g. for the runtime of the $(\mu + 1)$-EA on simple pseudo-boolean functions [160]. Other papers in this field concentrate on parameter control techniques, e.g. Jägersküpper's analysis [63], [64] of an EA with Rechenberg's 1/5th rule [114].

## 2.5  Summary

1. EAs are the attempt to adapt the powerful Darwinian evolution to computer science for search and optimization problems. They simulate the principles of inheritance, variation and natural selection with genetic operators. But this bloomy description means no more than EAs perform purely randomized search in the domain space.
2. Various operators and algorithmic schemes have been proposed, a few of them more, many less meaningful. The famous EA variants GAs, ES and EP have more and more grown together and cannot be distinguished anymore. This book concentrates on ES, which are appropriate for numerical optimization problems.
3. PSOAs exhibit similar elements like EAs, in particular ES. Operators like mutation, crossover and selection can be found in the PSOA equations.
4. Theoretical research comprises the fields no free lunch, Holland's schema theorem, the building block hypothesis, various dynamical systems approaches and statistical mechanics analysis. Very successful is the analysis of EAs from the point of view of Markov chains. In the last years, a classical runtime analysis of EAs got more and more attention.
5. The contribution of research in EA can be seen as old-fashioned AI research: reducing exponentially large search spaces by appropriate representations, operators and heuristics. The success of the search is often a result of the engineer's knowledge, also an old-fashioned AI wisdom and a symptom of the no free lunch theorem.
6. EAs have successfully been applied to domains like control, pattern recognition or automatic design. Sometimes the practitioner should be aware that classic search techniques may be faster or offer other advantages. But nevertheless, randomized search has grown to a strong, practicable and well-understood search technique, which is used all over the world in research and engineering.

# 3 Self-Adaptation

The adjustment of parameters and adaptive operator features is of crucial importance for reliable results to given problems and the efficiency of the evolutionary heuristics. But furthermore, proper parameter settings are important for the comparison of different algorithms on given problems. How can evolutionary parameters be controlled, how can they be tuned? This chapter gives an insight into parameter adaptation techniques. Beginning with the history of adaptation in EA, an extended taxonomy of parameter setting techniques is presented. Typically adapted components of EAs are presented. Afterwards, the chapter defines self-adaptation in a classic way and from a point of view of estimation of distribution algorithms, i.e. it gives insight into generalized views on self-adaptation and its limitations.

## 3.1 History of Parameter Adaptation

Historically, the development of adaptation mechanisms began in 1967, when Reed, Toombs and Baricelli [116] learned playing poker with an EA. The genome contained strategy parameters determining probabilities for mutation and crossover with other strategies. In the same year, Rosenberg [119] proposed to adapt the probability for applying crossover.

With the 1/5th rule Rechenberg [114] introduced an adaptation mechanism for step size control of ES, see section 3.2.2. One of the earliest realization of self-adaptation can be found in the work of Bagley [4], who was the first integrating the control parameters into the representation of the individuals. Today, the term self-adaptation is commonly associated with the self-adaptation of mutative step sizes for ES and was introduced by Schwefel [132] in 1974. For numerical representations the self-adaptation of mutation parameters seems to be an omnipresent feature. After ES, Fogel introduced self-adaptation to EP [43]. However, for binary-coded EAs self-adaptation has not grown to a standard method. Nevertheless, there are several approaches to incorporate it into binary representations, e.g. by Bäck [6], [7], Smith and Fogarty [144], Smith [142] and Stone and Smith [148]. Schaffer and Morishima [128] let the number and location

O. Kramer: Self-Adaptive Heuristics for Evolutionary Computation, SCI 147, pp. 29–47, 2008.
springerlink.com

of crossover points be controlled self-adaptively in their approach called *punctuated crossover*. Ostermeier et al. [102] introduced the cumulative path-length control, an approach to derandomize the adaptation of strategy parameters. Two algorithmic variants were the results of their attempts, the cumulative step-size adaptation (CSA) and the covariance matrix adaptation (CMA), also see section 4.1.6. Weinberg [157] as well as Mercer and Sampson [90] were the first who introduced metaevolutionary approaches. In metaevolutionary methods an outer EA controls the parameters of an inner one which optimizes the problem itself.

## 3.2  An Extended Taxonomy of Parameter Setting Techniques

After the design of the problem representation and the evaluation function the EA has to be specified. Besides the chose of the genetic operators, adequate parameters for the features of the EA have to be set. The parameter values determine the effectivity and the efficiency of the meta-heuristic. Appropriate operators have to be chosen as well as appropriate initializations of the strategy parameters. Furthermore, it can be useful to change these parameters during the run of the EA. In order to classify parameter tuning and control techniques of EAs the following aspects can be taken into account according to Eiben et al. [37]: What is changed, how is the change made, what is the scope/level of change and what is the evidence upon which the change is carried out.

### 3.2.1  Preliminary Work

Two main taxonomies of parameter adaptation have been introduced. Angeline [1] divides parameter adaptation schemes into concepts with *absolute update rules* and *empirical update rules*. Algorithms using *absolute update rules* change their parameters using fixed rules and statistical data about the population or over generations. The famous 1/5th success rule by Rechenberg [114] is an example for this class of parameter adaptation. Algorithms with *empirical update rules* control their parameters themselves. Strategy parameters are incorporated into the individuals' genome and are subject to crossover and mutation. The EA is able to control its parameters by letting individuals with high fitness values survive and inherit their strategy values. Furthermore, Angeline classifies *population-level, individual-level* and *component-level* parameters. *Population-level* parameters are changed for all individuals globally. *Individual-level* parameters are changed only for one individual while *component-level* adaptive methods affect each component of an individual separately.

Eiben, Hinterding and Michalewicz [37] extended Angeline's adaptation taxonomy by taking the *type* of adaptation and the *level* of adaptation into account. They distinguish between the following types of adaptation: *static* (user defined) parameter settings, *dynamic* parameter control, *adaptive* parameter control and *self-adaptive* parameter control. We take Eiben's taxonomy as the basis for our extensions.

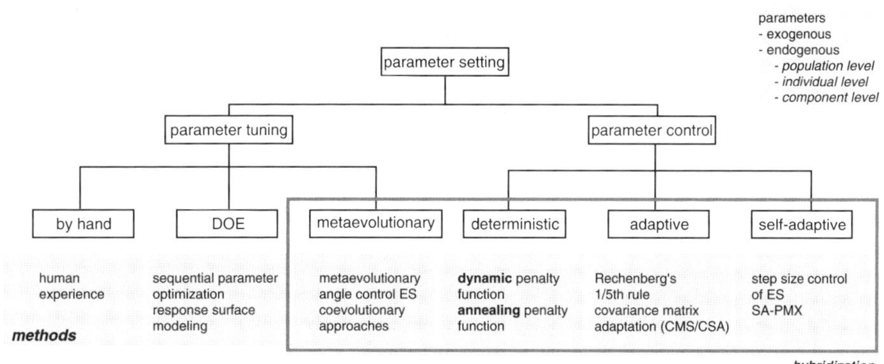


**Fig. 3.1.** An extended taxonomy of parameter setting techniques based on the taxonomy of Eiben [37], complemented on the parameter tuning branch. For each parameter setting class a couple of methods are presented exemplarily.

## 3.2.2   Parameter Tuning

Many parameters for EAs are static, which means that once defined they are not changed during the optimization process. Typical examples for static parameters are population sizes or initial strategy parameter values. Examples for static parameters of heuristic extensions are static penalty functions. In the approach of Homaifar, Lai and Qi [60], the search domain is divided into a fixed number of areas assigned with static penalties representing the constraint violation. The disadvantage of setting the parameters statically is the lack of flexibility concerning useful adaptations during the search process.

### By Hand

In most cases static parameters are set by hand. Hence, their influence on the behavior of the EA depends on human experience. Even though the user defined settings might not be the optimal ones.

### Design of Experiments, Relevance Estimation and Value Calibration

Design of experiments (DOE) offers the practitioner a way of determining optimal parameter settings. It starts with the determination of the objectives of an experiment and the selection of the parameters (factors) for the study. The quality of the experiment (response) guides the search to find appropriate settings. Recently Bartz-Beielstein et al. [5] developed a parameter tuning method for stochastically disturbed algorithm output, the sequential parameter optimization (SPO). It combines classical regression methods and statistical approaches for deterministic algorithms like design and analysis of computer experiments.

We conducted an experimental analysis of a particle swarm optimizer and ES using design of experiments [73].

Nannen and Eiben [97], [98], [31] proposed the relevance estimation and value calibration (REVAC) method to estimate the sensitivity of parameters and the choice of their values. It is based on information theory, i.e. REVAC estimates the expected performance when parameter values are chosen from a probability density distribution $C$ with maximized Shannon entropy. Hence, REVAC is an estimation of distribution algorithm. It iteratively refines a joint distribution $C$ over possible parameter vectors beginning with a uniform distribution giving an increasing probability to enlarge the expected performance of the underlying EA. From a vertical perspective new distributions for each parameter are built on estimates of the response surface, i.e. the fitness landscape. From a horizontal point of view in the first step the parameter vectors are evaluated according to the EA's performance, in the second step new parameter vectors are generated with superior response.

### Metaevolution

In metaevolutionary algorithms or nested EAs the evolutionary optimization process takes place on two different levels [115]. The outer EA optimizes parameters of the inner EA. When other adaptation methods such as adaptive or self-adaptive parameter control fail, nested EAs can be used to overcome the parameter control problem. Furthermore, nested EAs can be used for the experimental analysis of appropriate parameter settings. The PhD thesis of Kursawe [80] gives an overview of methods using nested or related approaches. He analyzed appropriate ES parameter settings, e.g. for the mutation parameter $\tau$, with a nested ES. Coello [24] makes use of nested ES for adapting the factors of a penalty function for constrained problems.

Figure 3.2 shows the pseudocode of a nested $[\mu'/\rho \overset{+}{,} \lambda'(\mu/\rho \overset{+}{,} \lambda)^\gamma]$-ES. In this algorithm $\mu'$ parents represent the settings for the inner $(\mu/\rho \overset{+}{,} \lambda)$-ES and reproduce $\lambda'$ offspring solutions. Each $(\mu/\rho \overset{+}{,} \lambda)$-ES runs for $\gamma$ generations, denoted as *isolation time*. The best $\mu'$ settings form the parents for the next generation on the outer level.

### 3.2.3  Parameter Control

Parameter control means the change of evolutionary parameters during the run. Typical is the control of strategy parameters, but also classical exogenous parameters like population sizes can be controlled during the run.

### Deterministic

Deterministic parameter control means that the parameters are adjusted according to a fixed time scheme. Usually, the number of generations controls the deterministic parameter control. E.g. it can be useful to reduce the mutation strength during the evolutionary search in order to enable convergence of the

| | |
|---|---|
| 1 | **Start** |
| 2 | $t_o := 0;$ |
| 3 | Initialize $\mathcal{P}_o(t);$ |
| 4 | Evaluate $\mathcal{P}_o(t);$ |
| 5 | **Repeat** |
| 6 | Change $\mathcal{P}_o(t) \to \mathcal{P}'_o(t)$ |
| 7 | Evaluate $\mathcal{P}'_o(t)$ by running the following subprogram; |
| 1 | **Start** |
| 2 | $t_i := 0;$ |
| 3 | Initialize $\mathcal{P}_i(t);$ |
| 4 | Evaluate $\mathcal{P}_i(t);$ |
| 5 | **Repeat** |
| 6 | Change $\mathcal{P}_i(t) \to \mathcal{P}'_i(t)$ |
| 7 | Evaluate $\mathcal{P}'_i(t);$ |
| 8 | Select $\mathcal{P}_i(t+1)$ from $\mathcal{P}'_i(t);$ |
| 9 | $t_i := t_i + 1;$ |
| 10 | **Until** termination condition |
| 11 | **End** |
| 8 | Select $\mathcal{P}_i(t+1)$ from $\mathcal{P}'_i(t);$ |
| 9 | $t_o := t_o + 1;$ |
| 10 | **Until** termination condition |
| 11 | **End** |

**Fig. 3.2.** Pseudocode of a nested EA

population. In the case of dynamic penalty functions the penalties are increased depending on the number of generations in order to avoid infeasible solutions in later phases of the search. Joines and Houck [66] propose the following dynamic penalty function:

$$\tilde{f}(\mathbf{x}) = f(\mathbf{x}) + (C \cdot t)^\alpha \cdot G(\mathbf{x}) \tag{3.1}$$

with fitness $f(\mathbf{x})$ of individual $\mathbf{x}$, the constraint violation measure $G(\mathbf{x})$ and parameters $C$ and $\alpha$. Annealing penalties are increased according to an external cooling scheme.

## Adaptive

Adaptive parameter control means the control of parameters according to a user defined heuristic. A feedback from the search is used to determine magnitude and direction of the parameter change. Furthermore, the heuristic defines whether its parameter values persist or propagate throughout the population. Adaptive penalty functions control the penalty of infeasible solutions according to a set of rules.

An example for an adaptive control of endogenous strategy parameters is the 1/5th success rule for ES by Rechenberg [114]. Running a simple (1+1)-ES with isotropic Gaussian mutations with constant mutation steps $\sigma$ and a simple objective function, the algorithm's strategy will become very slow after a certain number of generations, also see figure 3.5. The step sizes have to be adapted in order to speed up the optimization process. In the approach of the 1/5th-rule the whole populations makes use of the same $\sigma$ for all individuals. It states that the ratio of successful mutations should be 1/5th. If the ratio is greater than 1/5th the step size should be increased, if it is less than 1/5th the step size should be decreased. This rule is applied every $g$ generations. Figure 3.3 shows the pseudocode of Rechenberg's adaptation method. It must hold $0 < \lambda < 1$. The

1. Perform the (1+1)-ES for a number g of generations:

   - Keep $\sigma$ constant during this period,
   - Count the number $s$ of successful mutations during this period

2. Estimate success rate $p_s$ by

$$p_s := s/g \tag{3.2}$$

3. Change $\sigma$ according to

$$\sigma := \begin{cases} \sigma/\lambda, & \text{if } p_s > 1/5 \\ \sigma \cdot \lambda, & \text{if } p_s < 1/5 \\ \sigma, & \text{if } p_s = 1/5 \end{cases} \tag{3.3}$$

4. Go to step 1.

**Fig. 3.3.** Pseudocode of the 1/5th rule of Rechenberg

aim of this approach is to stay in the so called *evolution window* guaranteeing nearly optimal progress. Of course, the 1/5th rule is not the proper heuristic for every problem and is outperformed by self-adaptive step size approaches. The optimal value for the factor $a$ depends on several factors such as the number of generations g and the dimension of the problem $N$.

## Self-Adaptive

The concept of self-adaptation is a step into the direction of parameter independent EC. Self-adaptation is usually associated with the step size control of ES. It was originally introduced by Rechenberg and Schwefel [132], later by Fogel for EP. Self-adaptive algorithms are characterized by the integration of strategy parameters which influence the genetic operators into the individuals' chromosomes. These parameters undergo genetic variation and are therefore involved into the evolutionary optimization process

## Hybridization

The hybridization of different parameter control techniques is a source for many interesting variants. An example is the DSES, a heuristic for constraint handling, see chapter 7.4. The self-adaptive mutation strength control of ES fails in the vicinity of the infeasible search space. Hence, the adaptive heuristic takes control of the mutation strength. A minimum step size bounds the self-adaptive step sizes. A special heuristic cares for the reduction of this minimum step size in order to approximate the optimum.

## 3.3   Typical Adapted Parameters

This section summarizes parameters and features which are typically adapted. These reach from the problem representation and the fitness function to the genetic operators and the population. In the past research mainly focused on the parameter adaptation of mutation parameters, evaluation functions and population parameters.

### Representation

In classical GAs and ES the data structure is not subject to a change during the run of the algorithm. This is different for GP and classical EP where GP trees and finite state machines can vary in size and shape. Changing representations also occur in the delta encoding algorithm of Whitley and Mathias [158], which changes the representation of the function parameters in order to balance between exploration and exploitation.

### Fitness Evaluation Function

The fitness evaluation function is usually not changed during the run of an optimization algorithm as it is part of the problem itself[1]. An exception are the penalty functions for constrained problems. They make use of penalty factors which decrease the fitness of solutions according to their constraint violation. Penalty factors cannot be controlled self-adaptively because self-adaptation depends on selection and fitness evaluation. Individuals would tune their penalty factors in order to increase their fitness and selection probability respectively.

### Mutation Operators

Mutation operators are frequently subject to automatic parameter control. There are many conflicting results concerning the mutation probability of GAs. De Jong's [67] recommendation is $p_m = 0.001$, Schaffer et al. [127] recommend $0.005 \leq r_m \leq 0.01$ and Grefenstette [51] recommends $p_m = 0.01$. Mühlenbein [93] suggests to set the mutation probability $p_m = 1/l$ depending on the length $l$ of the representation. This discussion seems to be superfluous as the mutation strength is problem-dependent. The idea appears to control the mutation rate during the optimization run. Fogarty [41] suggests a deterministic control scheme

---

[1] This is not the case for dynamic optimization problems, where the problem changes over time.

which decreases the mutation strength depending on the number of generations. Bäck [7] decreases the mutation rate according to the distance to the optimum.

The most famous example is the step size of the ES Gaussian mutations, see chapters 2.2, 3.4 and 4.1. Bäck [7], [6] introduced a self-adaptive control mechanism for GA bit-string representations. But other deterministic and adaptive control mechanisms exist as well. For example the 1/5th rule of Rechenberg [114]. Behind mutation strength control lies the notion of exploring the search space with big mutations at the beginning of the search and afterwards exploiting the vicinity of the optimum in the exploitation phase with smaller variations. Chapter 4 gives a brief outline of self-adaptive mutation operators, e.g. Gaussian mutation of ES, correlated mutation or the control of parameters that bias the mutation into beneficial directions.

### Crossover Operators

Davis [30] adapts the crossover rates of his GA by rewarding crossover operators which created successful offspring. The approach makes use of different kinds of crossover operators. A reward increases the relative application probability of a successful crossover in comparison to the others. Schaffer and Morishima [128] introduced the punctuated crossover which adapts the crossover points for bit string representations. In chapter 6 various self-adaptive extensions of frequently used crossover operators are introduced.

### Selection

Most selection algorithms exhibit parameters which tune the selection pressure, e.g. the Boltzmann or ranking selection. Consequently, also these parameters can be subject to control mechanisms. The tournament size of tournament selection is another example for a tunable selection parameter. In the past, the control of the selection pressure has not been considered very often and is subject to current research. Eiben [35] introduced the self-adaptive control for tournament selection, also see section 3.5.

### Population

The population sizes of EAs influence the exploration and exploitation aspect of the evolutionary search. Many research activities have been investigated into appropriate populations sizes [37]. E.g. Smith [145] adjusts the population size according to the probability of a defined selection error. The approach of Hinterding et al. [56] controls the population sizes of three in parallel evolved populations according to heuristic rules. Arabas et al. [2] adjust the population size indirectly by assigning each individual with a maximum lifetime depending on its fitness. Furthermore, the expected number of offspring is proportional to the number of survived generations. Recently, Eiben [35] proposed an approach to control population sizes and selection pressure self-adaptively, as already mentioned. This approach is described in section 3.5.

## 3.4   The Concept of Self-Adaptation

Self-adaptation is usually associated with the step size control of ES. It was originally introduced by Rechenberg and Schwefel [132], later by Fogel [43] for EP. Self-adaptive algorithms are characterized by the integration of strategy parameters, which influence the genetic operators or other parameters of the EA, into the individuals' chromosomes. Typical is the influence of strategy parameters on mutation rates, crossover probabilities and recently population sizes. The chromosome of each individual $\mathbf{a} = (O, \Sigma)$ contains a set of solutions encoding objective variables $O$ and a set of strategy variables $\Sigma$. The strategy parameters $\sigma_i \in \Sigma$ are bound to the objective variables and evolved by means of the genetic operators recombination, mutation and selection. We summarize the concept of self-adaptation in the following definition.

**Definition 3.1 (Self-Adaptation of EA Parameters)**
*Self-adaptation is the evolutionary control of strategy parameters $\Sigma$, bound to each individual $\mathbf{a}$ and used to control parameters of the EA.*

Appropriate parameter settings have a positive effect on the individual's fitness and are consequently more likely inherited to the offspring. Hence, self-adaptation means an implicit control of the strategy parameters. Self-adaptation can be seen as a dynamical optimization process optimizing the strategy parameters during the walk through the search space. The *convergence* of the strategy parameters is a sign for a *good* working self-adaptive process, but it is no *necessary* condition.

In ES and EP the self-adaptive strategy variables are usually parameterizations of the mutation distributions. For example, an individual $\mathbf{a}$ of a $(\mu \overset{+}{,} \lambda)$-ES with objective variable vector $\mathbf{x}$ is mutated in the following way:

$$\mathbf{x}' := \mathbf{x} + \mathbf{z} \tag{3.4}$$

and

$$\mathbf{z} := (\sigma_1 \mathcal{N}_1(0, 1), \dots, \sigma_N \mathcal{N}_N(0, 1)) \tag{3.5}$$

The strategy parameter vector undergoes mutation

$$\boldsymbol{\sigma}' := e^{(\tau_0 \mathcal{N}_0(0,1))} \cdot \left( \sigma_1 e^{(\tau_1 \mathcal{N}_1(0,1))}, \dots, \sigma_N e^{(\tau_1 \mathcal{N}_N(0,1))} \right). \tag{3.6}$$

As the mutation step size has an important impact on the quality of the mutations and undergoes mutation itself, the evolutionary search controls the step size implicitly. Of course, self-adaptation is not restricted to steps sizes. Self-adaptive parameters also control the skewness of mutation probability functions, see chapter 4 or the probabilities of crossover.

Figure 3.4 illustrates the concept of self-adaptation. The individual performs the search in two subspaces: the objective and the strategy parameter search

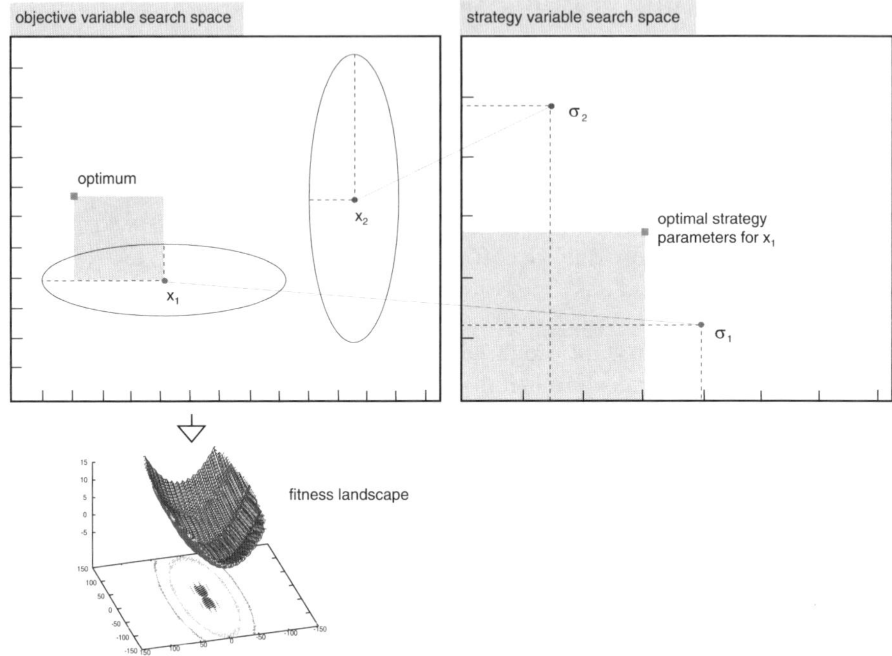


**Fig. 3.4.** The principle of self-adaptation. The figure illustrates the concept of searching in the objective and the strategy variable search space. We consider step sizes in a continuous search domain as an example. The strategy variable search space is displayed in double scale for better readability. For $x_1$ the yellow rectangular represents the *optimal* step sizes.

space. Strategy parameters influence the genetic operators of the objective variable space. The optimal settings vary depending on the location of the solution in the fitness landscape. Only the objective variables define the solution and have an impact on the fitness. The genetic operators for strategy and objective variables may vary, e.g. some paper propose dominant recombination and log-normal mutation for the strategy variables and intermediate and meta-EP mutation for the objective variables [18]. We can also think of decoupling strategy from objective variables and use different kinds of selection operators.

## Experimental Example

To give an example for the benefits of self-adaptation we compare a (15,100)-ES with a fixed mutation strength to a (15,100)-ES with self-adaptive mutation strength control, see equations 3.4 to 3.6, on a 2- and on a 30-dimensional sphere function. For the variants with a fixed mutation strength we set $\sigma = 0.001$ and

**Table 3.1.** Experimental comparison of an ES with a fixed step size on the 2-dimensional sphere function and the self-adaptive step size control (SA-ES) on a 30-dimension sphere function. The SA-ES shows outstanding results in comparison to the ES with constant step sizes.

| Experimental settings | |
|---|---|
| Population model | (15,100) |
| Mutation type | standard, $n_\sigma = N$, $\tau_0 = (\sqrt{2n})^{-1}$ and $\tau = (\sqrt{2\sqrt{n}})^{-1}$ |
| Crossover type | intermediate, $\rho = 2$ |
| Selection type | comma |
| Initialization | [-100;100] |
| Termination | 1000 generations |

| | best | worst | mean | dev |
|---|---|---|---|---|
| ES $\sigma = 0.1$ | 2.77E-05 | 2.55E-03 | 9.72E-04 | 0.00058 |
| ES $\sigma = 0.001$ | 3.57E-08 | 1.11E+02 | 1.20E+01 | 25.30838 |
| SA-ES | 2.92E-56 | 3.71E-50 | 3.36E-51 | 8.27E-51 |

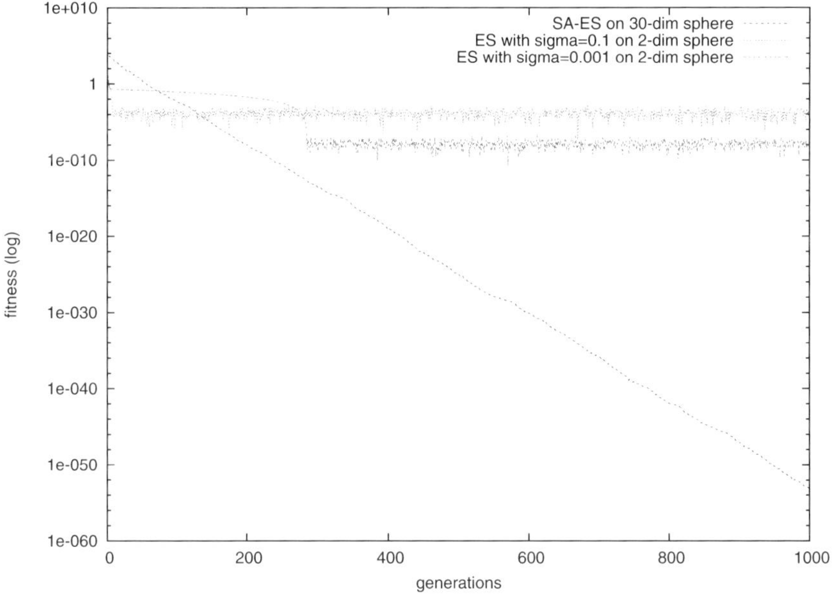


**Fig. 3.5.** Plot of typical runs of the variants ES with constant mutation strength ($\sigma = 0.1$ and $\sigma = 0.001$) and the self-adaptive ES on the sphere function. The approximation abilities of the self-adaptive process are outstanding in comparison to the ES with constant step sizes. The approximation abilities of a SA-ES on the 2-dimensional sphere function - not displayed here - go beyond the accuracy of the data structure after a few hundred generations.

$\sigma = 0.1$. For the self-adaptive algorithm we use the standard settings. Table 3.1 shows the outcome of the three experiments.[2]

The accuracy that can be achieved with the fixed mutation strengths is limited. The probability to approximate the optimum decreases as small mutations are not likely when the distance of the individuals to the optimum reach the degree of the step sizes. Figure 3.5 approves that the approximation capabilities with a fixed mutation strength are limited. Furthermore, the figure shows that the ES with $\sigma = 0.1$ is able to take bigger steps at the beginning whereas the ES with $\sigma = 0.001$ is rather slow. Otherwise, the latter is able to achieve a slightly better approximation as the mutation strength allows smaller mutations close to the optimum at the end of the search. The fitness with a fixed step size is fluctuating when reaching the area of the optimum within the range of the step size.

## 3.5   Self-Adaptation of Global Parameters

Eiben, Schut and Wilde [35] propose a method to control the population size and the selection pressure using tournament selection of an EA self-adaptively. The problem of such an adaptation is that the parameters *population size* and *tournament size* are global while self-adaptation is an approach for the individual or the component level. The idea of the approach is to derive the global parameters via aggregation of local information on individual or component level. The local information is still part of the individual's genome and participates in recombination and mutation. They introduce an aggregation mechanism which simply sums up all votes of the individuals for the global parameter. To the best of our knowledge this has been the only attempt to control global parameters self-adaptively.

## 3.6   Theoretical Approaches Toward Self-Adaptation

Only few theoretical work on self-adaptation exists. Most of it concerns the continuous search domain, i.e. the analysis of the ES mutation strength self-adaptation. As Beyer and Schwefel [18] state, the analysis of EAs including the mutation control part is a difficult task. In the following we give an overview over the main results.

### Dynamical Systems

Beyer [16] gives a detailed analysis of the $(1, \lambda)-$self-adaptation on the sphere function in terms of progress rate and self-adaptation response also considering the dynamics with fluctuations. The analysis reveals various characteristics

---

[2] We present the experimental results in this work by stating the best, the worst, the average, sometimes also the median result and the standard deviation (dev) of all runs. A result of a run is the best fitness within a population in the *last* generation. Graphical presentations like plots show either *typical* or *average* runs.

of the SA-$(1, \lambda)$–ES, e.g. it confirms that the choice for the learning parameter $\tau = c/\sqrt{N}$, see section 4.1.2, is reasonable. Beyer and Meyer-Nieberg [17] present results for $\sigma$-self-adaptation on the sharp ridge for a $(1, \lambda)$-ES without recombination. They use the evolution equation approximation and show that the radial and the axial progress rate as well as the self-adaptation response depend on the distance to the ridge axis. A further analysis reveals that the ES reaches a stationary normalized mutation strength. According to their analysis the function parameter $d$ (see equation 4.50) determines whether the ES concentrates on decreasing the distance to the ridge axis or increases the speed traveling along the ridge axis. If $d$ is smaller than a population size dependent limit the stationary step size leads to an increase of the distance to the ridge.

## Convergence Theory with Markov Chains and Supermartingales

Auger [3] investigated the $(1, \lambda)$–SA-ES on the one-dimensional sphere function. She proved sufficient conditions on the algorithm's parameters and showed that the convergence is $\frac{1}{t} \ln(\|X_t\|)$ with parent $X_t$ at generation $t$. Her proof technique is based on Markov chains for continuous search spaced and the Foster-Lyapunov drift conditions which enable stability property proofs of Markov chains. The main result is the log-linear type of the convergence and an estimation of the convergence rate. Even though the proof is very sophisticated, it does not show how the number of steps scales depending on the search space dimension.

The work of Semenov and Terkel [137] is based on a stochastic Lyapunov function. A supermartingale (stochastic process with a relationship between a random variable and its conditional expectation) is used to investigate the convergence of a simple self-adaptive EA. A central result of their research is that the convergence velocity of the analyzed self-adaptive EA is asymptotically exponential. They use Monte-Carlo simulations to validate the confidence of the martingale inequalities numerically, because they cannot easily be proven.

## Premature Convergence

Rudolph [126] investigates the premature convergence of EAs using self-adaptive mutation strength control. He assumes that the (1+1)-EA is located in the vicinity $\mathcal{P}$ of a local solution with fitness $f = \epsilon$. He proves that the probability to move from a solution in this set $\mathcal{P}$ to a solution in a set of solutions $\mathcal{S}$ with a fitness less than an $f < \epsilon$ (which in particular contains the global optimum, the rest of the search domain exhibits fitness values $f > \epsilon$ ) is smaller than 1. This result even holds for an infinite time horizon. Hence, an EA can get stuck at a non-global optimum with a positive probability. Rudolph also proposes ideas to overcome this problem.

## Runtime Analysis

Rechenberg proved that the optimal step size of ES depends on the distance to the optimum and created the 1/5th adaptation rule [114]. Recently, Jägersküpper

[63] proved that the (1+1)-ES using Gaussian mutations adapted by the 1/5-rule needs a linear number of steps, i.e. $O(n)$, to halve the approximation error on the sphere function. The bound $O(n)$ holds with "overwhelming probability of $1 - e^{-\Omega(n^{1/3})}$". Later, Jägersküpper [64] proved the bound $O(n \cdot \lambda\sqrt{\ln \lambda})$ for the $(1 + \lambda)$-ES with Rechernberg's rule. He proposes a modified 1/5th rule for $\lambda$ descendants as the regular algorithm fails for more than a single isotropic mutation. To the best of our knowledge, a runtime analysis for a self-adaptive ES has not been conducted yet with the exception of Auger's Markov chain result.

## 3.7   A Self-Adaptive Estimation of Distribution View on EAs

In the following we present a more generalized view on self-adaptation, i.e. we define self-adaptation from the point of view of estimation of distribution algorithms (EDAs). EDAs estimate the optimum with probability distributions.

### 3.7.1   Preliminary Views on Self-Adaptation

Bäck [8] gave insight into a generalized view on self-adaptation. From his point of view self-adaptation is biasing the population distribution to appropriate regions of the search space controlling the transmission function (crossover and/or mutation) between parents and offspring. Beyer and Meyer-Nieberg [91] point out that this can happen explicitly or implicitly. We refer to *explicit self-adaptation* if strategy variables exist which control the operator's properties. In contrast, *implicit self-adaptation* means that the operators exhibit properties that allow biasing the population distribution without containing strategy parameters. Beyer and Deb [33], [34] showed that binary GAs with well-designed crossover operators exhibit self-adaptive behavior, although they do not contain any strategy variables. They point out that one- and n-point-crossover can be seen as self-adaptive mutation operators for binary representations as common bit positions of the parents are transferred to the offspring. Consequently, the diversity in the parental population controls the diversity of the offspring population. Also Glickman and Sycara [46] see *implicit* self-adaptation as a non-injective genotype-phenotype mapping with variations of the genome. It exhibits an influence on the transmission function, but does not influence the fitness. Furthermore, the success of self-adaptation depends on the population's diversity. The self-adaptation of the operator's parameters improve the *convergence speed*. The degree of diversity determines the *convergence reliability*.

Another point of view on self-adaptation is the following. The strategy part of the genome is neutral concerning genotype-phenotype mapping, Igel and Toussaint [61] express that all neutral parts of the genome give the EA the ability to "vary the search space distribution independent of the phenotypic variation". A genome part is called *neutral* if it does not cause an influence on the phenotype and on the fitness. As the an analog argument holds for self-adaptation.

It changes the exploration distribution. They showed that neutrality is not generally a disadvantage concerning the performance, although the search space is increased.

### 3.7.2  Estimation of Distribution Algorithms

EDAs do not represent the solutions explicitly like EAs, but with probability distributions [106], [86], [81]. At the beginning of the 90s the first papers emerged yielding into the direction of EDAs. Syswerda [151] introduced a GA that makes use of statistical information gathered from the population. An easy example for EDAs and bit string representations estimate the setting of each chromosome with a probability, e.g. being 1. Most continuous EDAs make use of Gaussian probabilistic density distributions. From these distributions EDAs sample new solutions, select the best new samples and estimate the new distribution of the next iteration. We distinguish between univariate EDAs with no interactions between the variables, bivariate EDAs with pairwise interactions and multivariate EDAs with interactions between more than two variables. Famous instances of EDAs are compact genetic algorithms, univariate marginal distribution algorithms and population based incremental learning.

In recent years EDAs for continuous optimization problems became more and more famous [19], [122], [105], [136]. They try to fit the Gaussian model based on the diversity of the population with the maximum likelihood method. But it turned out that scaling and adaptation of model parameters for continuous domains is more complicated than for discrete search spaces. In terms of EDAs the control of the distribution variance is called *adaptive variance scaling*. Not all step size parameter control techniques known for EAs, see chapter 4, are applicable or have been investigated in the past. Most adaptive variance scaling techniques are based on Rechenberg's 1/5th success rule [114]. Self-adaptation is not efficient for EDAs due to the necessarily high number of competing populations. The derandomized strategy parameter control of the CMA-ES, section 4.1.6, has not been applied to EDAs yet.

Continuous EDAs using maximum likelihood suffer from premature convergence [49], [164], [100], [50]. Improperly shaped covariance matrices result in bad success rates and premature shrinking scaling factors. This phenomenon occurs when the center of the Gaussian model is too far away from the optimum [50] or when the fitness landscape is statistically deceptive in global scale [164]. Various techniques have been proposed recently to overcome this problem [22], [49], [20].

### 3.7.3  Self-Adaptation: Evolving the Estimation of Distribution

The EDA approaches inspire an EDA view on self-adaptation. This interpretation takes two steps and is exemplified for the step size self-adaptation of ES in the following. First, the population of an EA may be viewed as a generating finite mixture density function [27] like in the EDA approach. A similar comparison has already been proposed by Rudolph [123] for massively parallel simulated annealing. The second step concerns the properties of the mixture density functions. An appropriate variation operator should be able to adapt the properties of the probability distribution. In the case of EDAs, rules are invented, e.g. for the variance scaling. *Self-adaptation* evolves the density properties by evolution. This leads to the following definition.

**Definition 3.2 (Self-Adaptation of EA Parameters, EDA-view)**
*Self-adaptation is the evolutionary control of properties $\mathcal{P} \in \Sigma$ of a mixture density function $\mathcal{M}$ estimating the problem's optimum.*

Basis of the optimization of the distribution features $\mathcal{P}$ is evolutionary search bound to the search on the objective optimization level. From this point of view, an ES is an estimation of a mixture density function $\mathcal{M}$, i.e. $\mu$ Gaussian density functions estimate the optimum's location. From this mixture density function $\mathcal{M}$, $\lambda$ samples are generated. Hence, the *variance scaling* itself is an evolutionary process. Its success depends on the variation operators in the strategy variable search space $\Sigma$, i.e. log-normal or meta-EP mutation. Strategy variables and in most cases also heuristic extensions of EAs (e.g. the DSES) estimate the distribution $\mathcal{M}$ of the optimum's location. The Gaussian step sizes of ES are eminently successful, because its parameter directly control a meaningful distribution parameter, the deviation of the Gaussian distribution.

### 3.7.4  Views on the Proposed Operators and Problems of This Work

We pointed out that from the point of view of EDAs, an ES is an EDA with a Gaussian based mixture distribution $\mathcal{M}$ with evolutionary variance scaling. The EDA-view on self-adaptation can be found in most of the concepts of this work:

- BMO: The BMO is a biased mutation operator for ES, see section 4.2. Its biased mutation vector $\boldsymbol{\xi}$ shifts the center of the Gaussian mutation into beneficial directions within the degree of the step sizes $\boldsymbol{\sigma}$. From the EDA point of view, the BMO provides a mixture distribution $\mathcal{M}$ consisting of $\mu$ Gaussian distributions with an additional degree of freedom, i.e. the shift $\boldsymbol{\xi}$ of their centers.
- DMO: Similar to the BMO the DMO, see section 4.3, makes use of a mixture distribution $\mathcal{M}$ consisting of $\mu$ Gaussian distribution. But the shift of the Gaussian distribution results from an adaptive rule, i.e. the normalized vector

$$\xi = \frac{\chi_{t+1} - \chi_t}{|\chi_{t+1} - \chi_t|} \tag{3.7}$$

  of two successive populations' centers $\chi_{t-1}$ and $\chi_t$. This *global* bias shifts the single parts of the mixture distribution $\mathcal{M}$ depending on the recent past.
- SA-Crossover: Self-adaptive crossover is an attempt to exploit structural information of parts (blocks) of the solution during the optimization process. From the EDA point of view self-adaptive recombination for ES controls the distances of particular distributions $\mathcal{M}_i$ with $1 \leq i \leq \mu$ of the mixture distribution $\mathcal{M}$.

- Active Constraints: At the constraint boundary, the infeasible solutions cut off the mixture distribution $\mathcal{M}$. Hence under certain conditions, successful samples of $\mathcal{M}$ move to the constraint boundary with decreasing step sizes $\sigma$. The estimation of distribution process is constricted due to misleading success rates, see section 7.3.
- DSES: The DSES constraint handling technique proposed in section 7.4 establishes a minimum step size $\epsilon$ in order to avoid premature convergence. The interpretation from the EDA point of view is simple: The goal is to maintain least variances for the distributions $\mathcal{M}_i$. Samples from the distributions do not approximate the constraint boundary, but can reach the vicinity of the optimum, see section 7.3.
- TSES: The TSES introduced in section 7.5 is a constraint handling method for active optima. The TSES makes use of various heuristic modifications in order to approximate the optimum from two directions, the feasible and the infeasible region. The integration of knowledge can also be explained from an EDA perspective.

Every knowledge[3] that is incorporated into an EA, no matter how represented, is biasing the probability distributions $\mathcal{M}_i$ into directions which are supposed to be beneficial. EA researchers and practitioners design variation operators to establish this bias. But on the other side the win of performance is a loss of universality at the same time, because the distribution biases are not advantageous or not applicable on all kinds of problems.

## 3.8   Premature Convergence

Premature convergence of the mutation strength belongs to the most frequent problems of self-adaptive ES. It depends on the initialization of the step sizes, the learning parameters, the number of successful mutations and features of the search space. As evolution rewards short term success the evolutionary process can get stuck in local optima and suffer from premature convergence. Premature convergence is a result of a premature mutation strength reduction or a decrease of variance in the population. The problem is well known and experimentally proved. Only few theoretical works concentrate on this phenomenon, e.g. from Rudolph [126]. Premature convergence also appears for adaptive step control mechanisms. In section 7.3 we analyze the success rate situation at the constraint boundary that frequently leads to premature convergence.

The following causes for premature convergence could be derived. Stone and Smith [148] came to the conclusion that low innovation rates [4] and high selection pressure result in low diversity. They investigated Smith's discrete

---

[3] In this context the term *knowledge* comprises every information that helps to improve the optimization process. This reaches from heuristic modifications of algorithms, local search and intelligent initialization to memetic approaches

[4] Low innovation rates are caused by variation operators that produce offspring not far away from their parents, e.g. by low mutation rates.

self-adaptation GA on multimodal functions. Another reason for premature convergence was revealed by Liang et al. [85]. They point out that a solution with a high fitness but a far too small step size in one dimension is able to cause stagnation by inheriting this mutation strength to all descendants. As the mutation strength changes with a successful mutation according to the principle of self-adaptation, Liang et al. [85] considered the probability that after $k$ successful mutations the step size is smaller than an arbitrarily small positive number $\epsilon$. This results in a loss of step size control of their (1+1)-EP. Meyer-Nieberg and Beyer [91] point out that the reason for the premature convergence of the EP could be that the operators do not fulfill the postulated requirements for mutation operators, see section 4.1. Hansen [53] examined the conditions under which self-adaptation fails, in particular the inability of the step sizes to increase. He tries to answer the question whether a step size increase is affected by a bias of the genetic operators or due to the link between objective and strategy parameters. Furthermore, he identifies two properties of an EA: first, the descendants' object parameter vector should be point-symmetrically distributed after recombination and mutation. Second, the distribution of the strategy parameters given the object vectors after recombination and mutation should be identical for all symmetry pairs around the point-symmetric center.

In chapter 7 we prove the premature fitness reduction of a (1+1)-EA in the vicinity of the constraint boundary. It often leads to premature convergence. One way to overcome the premature convergence is the introduction of a lower bound for the mutation strength. The DSES of chapter 7 makes use of a lower bound. The disadvantage of such an approach is that the mutation strength also decreases during *convergence to the optimum*. But on the other hand the lower bound prevents the EA from convergence to the optimum due to too high and disturbing mutation strengths. In the DSES this shortcoming is handled by decreasing the lower bound adaptively with heuristic rules, i.e. based on the number of infeasible mutations at the constraint boundary.

## 3.9  Summary

1. EAs exhibit various parameters which have to be *tuned before* or which have to be *controlled during* the run. Parameter control can be classified into deterministic, adaptive, self-adaptive and metaevolutionary approaches. In deterministic approaches parameters change over the generations. Heuristic rules are the basis for adaptive parameter control.
2. Self-adaptation is an implicit parameter adaptation technique enabling the evolutionary search to tune the strategy parameters automatically. It is commonly used in ES and EP, mainly for the control of mutation parameters like the variance of the mutation distribution function. GAs in the classical form seldom use self-adaptation. In the past, self-adaptive crossover only focused on the adaptation of the crossover probability. Beyer and Meyer-Nieberg [91] point out that crossover in binary standard GAs exhibit a form of self-adaptation.

3. The analysis of self-adaptation either concentrates on simplified models of the algorithm or relies on numerical calculations. Theoretical research about self-adaptation can mainly be divided into approaches expressing the so-called evolution equations, the analysis of Markov chains or the theory of martingales.
4. The EDAs help to get a useful view on self-adaptation. It can be seen as the evolutionary control of properties $\mathcal{P} \in \Sigma$ of a mixture density function $\mathcal{M}$ estimating the optimum of the problem.
5. Premature convergence is the biggest drawback of self-adaptation. It mainly happens to mutation strength as a result of too high selection pressure and low diversity in the population.

# Part II

# Self-Adaptive Operators

# 4 Biased Mutation for Evolution Strategies

The Gaussian mutation operator of ES exhibits the feature of unbiasedness of its mutations. But offering a bias to particular directions can be a useful degree of freedom, in particular when controlled self-adaptively. This chapter introduces our biased mutation operator (BMO) [77], [78] with its self-adaptive bias control together with a couple of variants. Among the variants is the descent direction mutation operator (DMO), which is based on the descent direction between the center of two successive populations. We prove that biased Gaussian mutation is superior to unbiased mutation on monotone functions as long as the bias points into the descent direction. Various tests revealed successful results, in particular on ridge functions and in constrained search domains, but also on some multimodal problems. In the past, various self-adaptive mutation operators for ES have been proposed, i.e. for optimization in numerical search domains. They reach from self-adaptive uncorrelated isotropic Gaussian mutation [113], [131] to the derandomized step size control of the covariance matrix adaptation [54].

This chapter is structured as follows. At first, we recapitulate standard mutation operators and some well-known variants for ES in section 4.1. Afterwards, section 4.2 introduces the BMO, a self-adaptive BMO for ES, with the coefficient vector $\xi$ and it self-adaptation mechanism. We introduce a number of variants: the sphere BMO (sBMO) with $N$ bias coefficients but only one step size, a cube BMO (cBMO) with one bias coefficient and $N$ step sizes. The notion of a constant direction of a population on monotone function parts leads to the idea of adapting the bias according to successive populations' centers: the DMO adapts the bias according to the descent direction estimated by two successive populations' centers.

All variants are tested and evaluated on a set of test problems in section 4.5. The biased mutation achieves the greatest improvements on the tested ridge functions and on constrained problems. On multimodal problems slight improvements were achieved.

## 4.1 Mutation Operators for Evolution Strategies

For a comprehensive introduction to ES, we refer to chapter 2.2 and to Beyer and Schwefel [18]. Here it is important to keep in mind that an individual consists

O. Kramer: Self-Adaptive Heuristics for Evolutionary Computation, SCI 147, pp. 51–80, 2008.
springerlink.com

of a vector of objective and strategy variables. In real-valued search spaces the objective variables are real values $x_1, \ldots, x_N$ representing the assignment of the variables of an $N$-dimensional optimization problem. The strategy variables contain additional information, usually parameters for the Gaussian mutation operator. There are three main principles for the design of mutation operators proposed by Beyer [16]:

- reachability,
- unbiasedness, and
- scalability.

The first principle *reachability* ensures that the whole objective and strategy parameter search space can be reached within a finite number of generations. This is also a necessary condition for proving global convergence, also see our convergence proof for inversion mutation in section 5.3. The *scalability* condition ensures that the mutation strength can adapt to values, which guarantee improvements during the optimization process. The condition of *unbiasedness* is appropriate to many unconstrained real search spaces. But for constrained problems EAs with a self-adaptive step size mechanism often suffer from a disadvantageous success probability at the boundary of the feasible search space [75], also see chapter 7. For this problem class biased mutation is an appropriate technique, see section 4.5.3 of the this chapter.

Various mutation operators for ES exist. For the basic $(\mu \, {}^+_, \lambda)$-ES uncorrelated isotropic mutation was introduced by Rechenberg [113] and Schwefel [131] as well as correlated mutation by Schwefel [132]. Many extensions and variations

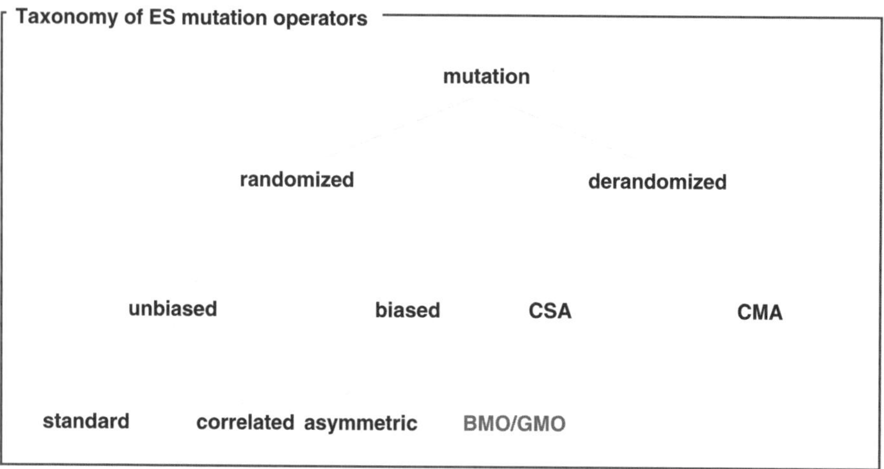


**Fig. 4.1.** A small taxonomy of mutation operators for ES. Mutation operators can be classified into randomized and derandomized mutation of the mutation strength. Within the category of randomized operators we distinguish between unbiased and biased mutation.

were introduced within the last years. Figure 4.1 gives an overview of the most famous mutation operators for ES. We distinguish between randomized and de-randomized step size mutation. Cumulative step size adaptation (CSA) by Ostermeier, Gawelczyk and Hansen [102] and covariance matrix adaptation (CMA) by Hansen [54] fall into the latter category. The randomized operators can be classified into biased and unbiased variants. The ES standard mutation has already been mentioned. Correlated mutation was proposed by Schwefel [132]. A directed mutation operator is the asymmetric mutation by Hildebrand [55]. Another variant is the BMO introduced in this work. Further operators are based on other kinds of distributions like the Cauchy distribution by Yauo and Liu [162].

### 4.1.1  Uncorrelated Isotropic Gaussian Mutation with One Step Size

Mutation is the main source for variation in the evolutionary process. In real-valued search spaces mutation means the addition or subtraction of a small variation, so that a child evolves with a high similarity to its parents. The uncorrelated isotropic mutation for ES uses only $n_\sigma = 1$ endogenous strategy variable $\sigma$, which represents the standard deviation of the normal distribution. For the sake of better understanding it can be seen as the radius of the sphere within mutations are created. For ES in real-valued search spaces objective variables are mutated by

$$\mathbf{x}' := \mathbf{x} + \mathbf{z} \tag{4.1}$$

with mutation

$$\mathbf{z} := \sigma(\mathcal{N}_1(0,1),\ldots,\mathcal{N}_N(0,1)), \tag{4.2}$$

where $\mathcal{N}_i(0,1)$ are independent random samples from the standard normal distribution with expected value 0 and standard deviation 1. The strategy variable itself is mutated using the log-normal rule

$$\sigma' := \sigma e^{(\tau\mathcal{N}(0,1))} \tag{4.3}$$

with the recommendation $\tau = \frac{1}{\sqrt{N}}$ for the *learning rate*. Depending on the Gaussian random value produced by $\tau\mathcal{N}(0,1)$ in the exponent of the $e$-function, the step size $\sigma'$ is decreased or increased. But with a high probability the change is small. The resulting distribution is a log-normal distribution. This mechanism is the key to self-adaptation of the step sizes. Each component $x'_i$ of mutant $\mathbf{x}'$ obeys the density function

$$p(x'_i) = \frac{1}{\sqrt{2\pi}\sigma} \exp\left(-\frac{1}{2}\frac{(x'_i - x_i)^2}{\sigma^2}\right). \tag{4.4}$$

Due to the symmetry of the distribution function the Gaussian mutation introduces no bias.

### 4.1.2  Uncorrelated Gaussian Mutation with $N$ Step Sizes

In the basic $(\mu \stackrel{+}{,} \lambda)$-ES usually a vector of $n_\sigma = N$ step sizes is used, which results in mutation ellipsoids:

$$\mathbf{z} := (\sigma_1 \mathcal{N}_1(0, 1), \dots, \sigma_N \mathcal{N}_N(0, 1)) \tag{4.5}$$

The corresponding strategy parameter vector is mutated with the extended log-normal rule:

$$\boldsymbol{\sigma}' := e^{(\tau_0 \mathcal{N}_0(0,1))} \cdot \left( \sigma_1 e^{(\tau_1 \mathcal{N}_1(0,1))}, \dots, \sigma_N e^{(\tau_1 \mathcal{N}_N(0,1))} \right) \tag{4.6}$$

The parameters $\tau_0$ and $\tau_1$ have to be tuned. Recommended settings are $\tau_0 = \frac{c}{\sqrt{2N}}$ and $\tau_1 = \frac{c}{\sqrt{2\sqrt{N}}}$ and $c = 1$ as a reasonable choice for a (10,100)-ES [18]. Comprising, an individual $\mathbf{a}$ consists of the object parameter set $x_i$ with $1 \leq i \leq N$, the mutation strength vector and the assigned fitness $F(\mathbf{x})$. So it is specified by

$$\mathbf{a} = (x_1, \dots, x_N, \sigma_1, \dots, \sigma_N, F(x)). \tag{4.7}$$

### 4.1.3  Correlated Mutation

For some fitness landscapes it is more beneficial to use a rotated mutation ellipsoid for the purpose of an improvement of the success rate. Rotation of the mutation ellipsoid is achieved by the correlated mutation proposed by Schwefel [132]. For an $N$-dimensional problem $k = N(N-1)/2$ additional strategy parameters, the angles for the rotation of the mutation ellipsoid, are introduced. Let $\sigma$ again be the vector of step sizes and $\mathbf{M}$ be the orthogonal rotation matrix. The mutations are produced in the following way:

$$\mathbf{z} := \mathbf{M}(\sigma_1 \mathcal{N}_1(0, 1), \dots, \sigma_N \mathcal{N}_N(0, 1)) \tag{4.8}$$

The probability density function for $\mathbf{z}$ becomes

$$p(\mathbf{z}) = \frac{e^{-\frac{1}{2}\mathbf{z}^T \cdot C^{-1} \cdot \mathbf{z}}}{(\det C \cdot (2\pi)^n)^{\frac{1}{2}}} \tag{4.9}$$

with the rotation matrix[1] $C$ and entries

$$c_{ii} = \sigma_i^2, \tag{4.10}$$

$$c_{ij, i \neq j} = \begin{cases} 0 & \text{no correlations,} \\ \frac{1}{2}(\sigma_i^2 - \sigma_j^2) \tan(2\alpha_{ij}) & \text{correlations.} \end{cases} \tag{4.11}$$

---

[1] The rotation matrix $C$ is also called *covariance matrix*.

For a mutation in two dimensions the vector of objective variables has to be mutated with the following trigonometric matrix

$$\begin{pmatrix} \cos(\alpha_{ij}) & -\sin(\alpha_{ij}) \\ \sin(\alpha_{ij}) & -\cos(\alpha_{ij}) \end{pmatrix}. \tag{4.12}$$

For a self-adaptation process the $k$ angles $\alpha_1, \ldots \alpha_k$ have to be mutated. Schwefel [132] proposed

$$\alpha' = \alpha + \gamma \cdot \mathcal{N}(0, 1) \tag{4.13}$$

with $\gamma = 0.0873$ corresponding to $5°$.


### 4.1.4  Asymmetric Density Functions - Directed Mutation

Similar to the idea of correlated mutation, the assumption of directed mutation is that in some parts of the landscape a success rate improvement can be achieved by skewing the search into a certain direction. Hildebrand [55] proposes an asymmetrical mutation operator. This approach demands an asymmetry parameter set of $N$ additional parameters $\mathbf{c} = (c_1, \ldots, c_N)$. These parameters determine the mutation direction and therefore only cause linear growth of the strategy parameters instead of the quadratic growth of correlated mutation. In order to decouple asymmetry from the step sizes, a normalized directed mutation operator has recently been proposed by Berlik [13]. The density function for the normalized directed mutation is the following:

$$f_{\sigma,a}(x) = \begin{cases} \dfrac{2}{\pi} \dfrac{\sqrt{1-a}}{(1+\sqrt{1-a})\sigma_{\mathrm{norm}}(a)\sigma} e^{-\frac{x^2}{2(\sigma(a)\sigma)^2}} & \text{for } a \leq 0, x \leq 0 \\[2ex] \dfrac{2}{\pi} \dfrac{\sqrt{1-a}}{(1+\sqrt{1-a})\sigma_{\mathrm{norm}}(a)\sigma} e^{-\frac{(1-a)x^2}{2(\sigma(a)\sigma)^2}} & \text{for } a \leq 0, x > 0 \\[2ex] \dfrac{2}{\pi} \dfrac{\sqrt{1+a}}{(1+\sqrt{1+a})\sigma_{\mathrm{norm}}(a)\sigma} e^{-\frac{(1+a)x^2}{2(\sigma(a)\sigma)^2}} & \text{for } a > 0, x \leq 0 \\[2ex] \dfrac{2}{\pi} \dfrac{\sqrt{1+a}}{(1+\sqrt{1+a})\sigma_{\mathrm{norm}}(a)\sigma} e^{-\frac{x^2}{2(\sigma(a)\sigma)^2}} & \text{for } a > 0, x > 0 \end{cases} \tag{4.14}$$

and its normalization function $\sigma_{\mathrm{norm}}(a)$. The generation of corresponding random numbers is a rather complicated undertaking. The idea of skewing the mutations into a certain direction is similar to our approach. But our approach is easier to implement. Figure 4.4 visualizes the effect of the directed mutation operator when mutations are skewed into a certain direction. The highest probability to reproduce mutations is still in the neighborhood environment of the child.

### 4.1.5  Cauchy Mutation

Instead of the Gaussian mutation the so called fast evolutionary programming (FEP) [163] makes use of the Cauchy distribution. The one-dimensional Cauchy density function is defined by

$$p(x) = \frac{1}{\pi} \frac{t}{t^2 + x^2}, \quad -\infty < x < \infty \tag{4.15}$$

with a scale parameter $t > 0$ [163]. The corresponding distribution is

$$\mathcal{P} = \frac{1}{2} + \frac{1}{\pi} \arctan\left(\frac{x}{t}\right). \tag{4.16}$$

The mutation of an individual $\mathbf{x}$ follows the principle of the uncorrelated Gaussian mutation:

$$\mathbf{x}' := \mathbf{x} + \mathbf{z} \tag{4.17}$$

with the mutation

$$\mathbf{z} := \sigma(\mathcal{P}_1(t), \dots, \mathcal{P}_N(t)) \tag{4.18}$$

in the case of one step size and the mutation

$$\mathbf{z} := (\sigma_1 \mathcal{P}_1(t), \dots, \sigma_N \mathcal{P}_N(t)) \tag{4.19}$$

in the case of $N$ step sizes.

### 4.1.6  Covariance Matrix Adaptation (CMA)

The covariance matrix adaptation evolution strategy (CMA-ES) by Hansen and Ostermeier [54] is a successful evolutionary optimization method for real-parameter optimization of non-linear functions. The basis of this approach is a derandomized step-size adaptation. The mutation distribution is altered deterministically so that the probability to reproduce steps in the search space, which have led to the current population, is increased. The CMA-ES makes use of a population of search points, which are sampled with a normal distribution, see equation 4.20. The idea of the CMA is similar to quasi-Newton [52]: approximation of the inverse Hessian matrix, i.e. fitting the search distribution to the objective function curvature. The complete algorithm and the full explanation of all symbols can be found in the tutorial of Hansen [52]. Here we shortly repeat the core algorithm that can also be found in the tutorial of Hansen [52], pages 22-24.

We use the following symbols:

- the sample of $\lambda$ search points $\mathbf{x}_k^{(t+1)} \in \mathbb{R}^N$ for $k = 1, \dots, \lambda$ of generation $t + 1$,

- the multivariate normal distribution $\mathcal{N}(\mathbf{m}, \mathbf{C})$ with mean $\mathbf{m}$ and covariance matrix $\mathbf{C}$,
- the $i$-th best point $\mathbf{x}_{i:\lambda}^{(t+1)}$ out of $\mathbf{x}_1^{(t+1)}, \ldots, \mathbf{x}_\lambda^{(t+1)}$ from equation 4.20,
- $\mu_{\text{eff}} = (\sum_{i=1}^\mu w_i^2)^{-1}$ is the variance effective selection mass, $1 \leq \mu_{\text{eff}} \leq \mu$,
- OP: $\mathbb{R}^N \rightarrow \mathbb{R}^{N \times N}, \mathbf{x} \mapsto \mathbf{x}\mathbf{x}^T$.

The CMA-ES works in the following steps.

- **Parameter setting.** Set $\lambda, \mu, w_{i=1\ldots\mu}, c_\sigma, d_\sigma, c_c, \mu_{\text{cov}}, c_{\text{cov}}$ to default values.
- **Initialization.** Set evolution path $\mathbf{p}_\sigma^{(0)} = \mathbf{0}, \mathbf{p}_c^{(0)} = \mathbf{0}$. Set covariance matrix $\mathbf{C}^{(0)} = \mathbf{I}$. Set initial step size $\sigma^{(0)} \in \mathbb{R}_+$, distribution mean $\mathbf{m}^{(0)} \in \mathbb{R}^n$ problem dependent.
- **Iteration loop.** $t = 0, 1 \ldots$ until termination criterion fulfilled.
  - *sample* new population of search points

$$\mathbf{x}_k^{(t+1)} \mathcal{N}(\mathbf{m}^{(t)}, (\sigma^{(t)})^2 \mathbf{C}^{(t)}) \text{ for } k = 1, \ldots, \lambda \qquad (4.20)$$

  - *selection and recombination*

$$\mathbf{m}^{(t+1)} = \sum_{i=1}^\mu w_i \mathbf{x}_{i:\lambda}^{(t+1)}, \text{ where } \sum_{i=1}^\mu w_i = 1, w_i > 0 \qquad (4.21)$$

  - *step size control*

$$\mathbf{p}_\sigma^{(t+1)} = (1 - c_\sigma)\mathbf{p}_\sigma^{(t)} + \sqrt{c_\sigma(2 - c_\sigma)\mu_{\text{eff}}} \mathbf{C}^{(t)-\frac{1}{2}} \frac{\mathbf{m}^{(t+1)} - \mathbf{m}^{(t)}}{\sigma^{(t)}} \qquad (4.22)$$

$$\sigma^{(t+1)} = \sigma^{(t)} \exp\left( \frac{c_\sigma}{d_\sigma} \left( \frac{||\mathbf{p}_\sigma^{(t+1)}||}{E||\mathcal{N}(\mathbf{0}, \mathbf{I})||} - 1 \right) \right) \qquad (4.23)$$

  - *covariance matrix adaptation*

$$\mathbf{p}_c^{(t+1)} = (1 - c_c)\mathbf{p}_c^{(t)} + h_\sigma^{(t+1)} \sqrt{c_c(2 - c_c)\mu_{\text{eff}}} \frac{\mathbf{m}^{(t+1)} - \mathbf{m}^{(t)}}{\sigma^{(t)}} \qquad (4.24)$$

$$\mathbf{C}^{(t+1)} = (1 - c_{\text{cov}})\mathbf{C}^{(t)} + \frac{c_{\text{cov}}}{\mu_{\text{cov}}} \left( \mathbf{p}_c^{(t+1)} \mathbf{p}_c^{(t+1)T} + \delta(h_\sigma^{(t+1)})\mathbf{C}^{(t)}) \right) \qquad (4.25)$$

$$+ c_{\text{cov}} \left( 1 - \frac{1}{\mu_{\text{cov}}} \right) \sum_{i=1}^\mu w_i \text{OP}\left( \frac{\mathbf{x}_{i:\lambda}^{(t+1)} - \mathbf{m}^{(t)}}{\sigma^{(t)}} \right) \qquad (4.26)$$

The CMA-ES works with a *global* step, the scaling factor and an *individual* step, the covariance matrix. The global step is computed considering the mean distance between the successful mutated offspring and the current population center, see equation 4.23. A further concept that is used is the cumulative path-length control by Ostermeier, Hansen and Gawelczyk [101]. The information of previous generations is used. Cumulative step size adaptation makes use of an evolution path with *cumulative path length* $||p||$, relative to the mean distance of selected offspring around the center, see equation 4.22.

### 4.1.7 Self-Adaptive Mutation for Binary Representations

From the point of view of GAs the mutation operator has always been seen as a background operator [59]. Traditionally, the main variation operator for GAs is the crossover operator. Beyer and Meyer-Nieberg point out that standard bit flip mutation for binary representations introduces a bias [91]. Self-adaptation for binary representations was introduced by Bäck [7], [6]. He encoded the mutation rate as a bit-string. The self-adaptation works as follows: the mutation rate is decoded to a value in the interval $[0, 1]$ and used to mutate its own bit-string representation. The mutation rate is applied to the bit-strings of the objective variables. An analysis of the asymptotic behavior neglecting recombination and selection showed that the mutation rate results in a Markov chain with the absorbing state zero. Bäck and Schütz [9] showed that the mutation rate encoded as a bit-string is counterproductive for self-adaptation. They introduced a real-coded mutation rate to overcome this problem. Smith and Fogarty [144] extended the work of Bäck and examined the behavior of a $(\mu + 1)$-GA.

### 4.1.8 Mutation Operators for Strategy Parameters

There are two main variants for the mutation of strategy parameters [16]: *Meta-EP* mutation has up to now mainly been applied to the angle adaptation of the correlated ES,

$$\Lambda' := \Lambda + \gamma \cdot \mathcal{N}(0, 1). \tag{4.27}$$

The exogenous parameter $\gamma$ is a learning parameter and defines the mutation strength. A second operator is the log-normal mutation operator by Schwefel [132]. It has already been introduced in this chapter.

$$\Lambda' := \Lambda \cdot e^{\gamma \cdot \mathcal{N}(0,1)} \tag{4.28}$$

In the case of step sizes of ES, $\gamma$ is called $\tau$. Variants like the extended log-normal mutation exist, see section 4.1.2. Rechenberg [115] introduced a discrete mutation operator using a symmetric two-point distribution, the two-point-operator

$$\Lambda' := \begin{cases} \Lambda(1 + \gamma) & \text{if } u(0, 1] \leq 1/2 \\ \Lambda/(1 + \gamma) & \text{if } u(0, 1] > 1/2. \end{cases} \tag{4.29}$$

Again, parameter $\gamma$ is a learning parameter, $u$ is a uniformly distributed random number.

## 4.2 The Biased Mutation Operator

In this section we introduce our BMO, starting with the main operator. The two variants sphere and cube BMO are introduced subsequently.

### 4.2.1   BMO Concept

Unlike directed mutation, the BMO does not change the skewness, but biases the mean of the Gaussian distribution to lead the search into a more beneficial direction. This is reflected in the success rate of reproducing superior offspring. For the BMO we introduce a *bias coefficient vector* $\boldsymbol{\xi}$ which indicates the level of bias relative to the standard deviation $\sigma$

$$\boldsymbol{\xi} = (\xi_1, \ldots, \xi_N) \text{ with } -1 \leq \xi_i \leq 1. \tag{4.30}$$

For every $i \in 1, \ldots, N$ the bias vector $\mathbf{b} = (b_1, \ldots, b_N)$ is defined by:

$$b_i = \xi_i \cdot \sigma_i \tag{4.31}$$

Since the absolute value of bias coefficient $\xi_i$ is less than or equal to 1, the bias will be bound to the step sizes $\sigma_i$. This restriction prevents the search from being biased too far away from the parent. Figure 4.2 illustrates the BMO. The BMO follows the standard way of mutation

$$\mathbf{x}' := \mathbf{x} + \mathbf{z}. \tag{4.32}$$

The mutation of the BMO works as follows:

$$\begin{aligned}
\mathbf{z} :=\ & (\sigma_1 \mathcal{N}_1(0,1) + b_1, \ldots, \sigma_N \mathcal{N}_N(0,1) + b_N) & (4.33)\\
=\ & (\sigma_1 \mathcal{N}_1(0,1) + \xi_1 \sigma_1, \ldots, \sigma_N \mathcal{N}_N(0,1) + \xi_N \sigma_N) & (4.34)\\
=\ & (\sigma_1 \mathcal{N}_1(\xi_1, 1), \ldots, \sigma_N \mathcal{N}_N(\xi_N, 1)) & (4.35)
\end{aligned}$$

In terms of modifying the mutation strength, the aforementioned log-normal rule is applied. Furthermore, in the BMO the bias coefficients are mutated in the following meta-EP way:

$$\xi_i' = \xi_i + \gamma \cdot \mathcal{N}(0,1) \quad i = 1, \ldots, N. \tag{4.36}$$

The parameter $\gamma$ is a new parameter introduced for the BMO to determine the mutation strength on the bias. In section 4.5 recommendations will be proposed to tune this parameter.

The BMO biases the mean of mutation and enables the ES to reproduce offspring outside the standard mutation ellipsoid. To direct the search, the BMO enables the mutation ellipsoid to move within the bounds of the regular step sizes $\sigma$. The bias moves the center of the Gaussian distribution within the bounds of the step sizes. This is advantageous on ridge functions, on some multimodal functions and at the edge of feasibility. Without the BMO the success rate to reproduce better offspring is relatively low because many mutations lie beyond the feasible search space or have got a worse fitness, as described in section 7.3.

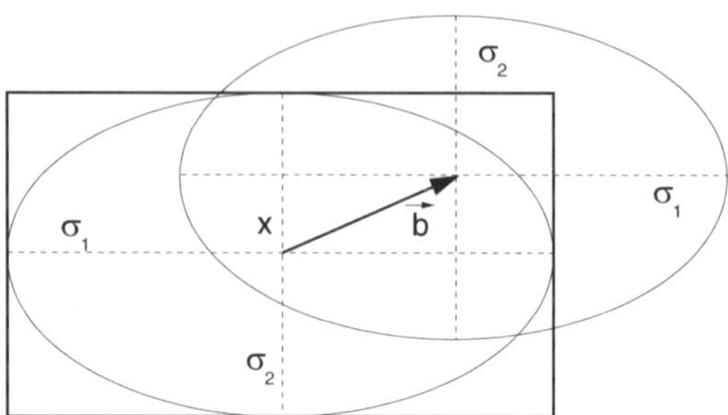

**Fig. 4.2.** Concept of the BMO in two dimensions: The center of the mutation ellipsoid is shifted by the bias coefficient vector **b** within the bounds of the step sizes $\boldsymbol{\sigma}$

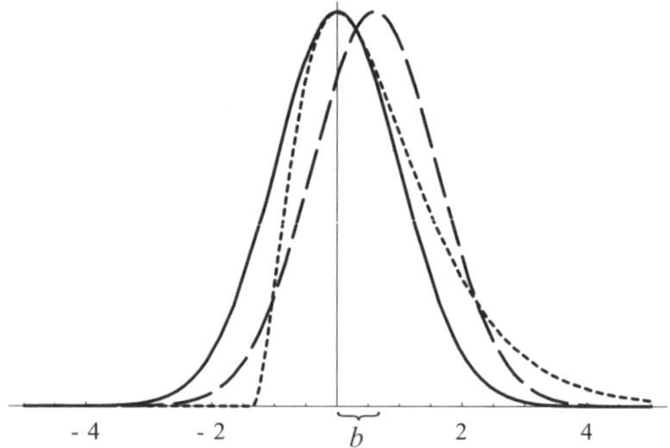

**Fig. 4.3.** Comparison of standard mutation (solid), directed mutation (dotted) and biased mutation (dashed) with bias $b$

The bias coefficient vector $\boldsymbol{\xi}$ improves the success rate situation as the success area increases. The BMO approach is as flexible as correlated and directed mutation, but is less computational expensive than both methods, see section 4.2.4. In comparison to the directed mutation the computation of the random numbers of an asymmetric probability density function is usually more computationally expensive than the computation of Gaussian random numbers. Especially, in practice the implementation is less complex. Figure 4.3 shows a comparison of the different probability density functions of the three mentioned approaches. In the following, we introduce various variants of the standard BMO. In section 4.5 their capabilities are examined experimentally.

## 4.2.2   Sphere Biased Mutation Operator (sBMO)

We propose the Sphere Biased Mutation Operator (sBMO) as a variant of the BMO with only one step size analog to the isotropic Gaussian mutation with one

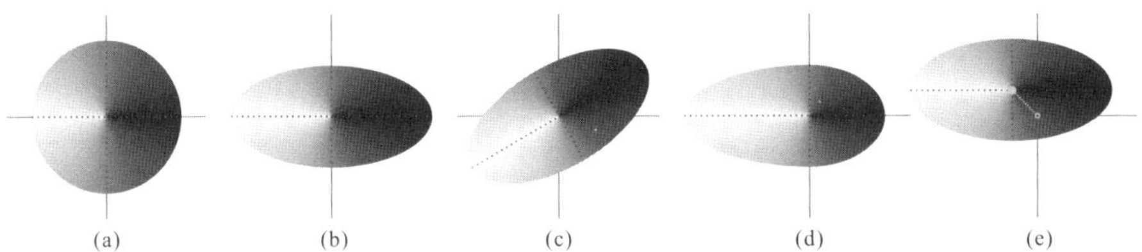

**Fig. 4.4.** The mutation ellipsoids of the different mutation operators in a two-dimensional search space: For the purpose of better understanding we assume all mutations fall into the ellipsoids instead of taking the probability density functions into account. From left to the right: (a) uncorrelated mutation with one step size, (b) uncorrelated mutation with $N = 2$ step sizes, (c) correlated mutation resulting in rotation, (d) directed mutation with different skewness and (e) biased mutation making use of the *bias coefficient vector*.

step size, but again with a $N$-dimensional bias coefficient vector $\boldsymbol{\xi} = (b_1, \ldots, b_N)$ with $b_i = \xi_i \cdot \sigma_i$. Hence, the mutation $\mathbf{z}$ becomes

$$\mathbf{z} := \sigma(\mathcal{N}_1(\xi_1, 1), \ldots, \mathcal{N}_N(\xi_N, 1)). \tag{4.37}$$

Mutations of the sBMO lie within a hypersphere. The bias $\boldsymbol{\xi}$ is mutated analog to the biased mutation of the standard BMO in order to enable self-adaptation, see equation 4.36.

### 4.2.3  Cube Biased Mutation Operator (cBMO)

Another variant of the BMO is the use of only one bias value $b$, which spans a hypercube with the edge length $b$, but $N$ step sizes. The mutation $\mathbf{z}$ becomes

$$\mathbf{z} := (\sigma_1 \mathcal{N}_1(b, 1), \ldots, \sigma_N \mathcal{N}_N(b, 1)). \tag{4.38}$$

Mutations of the cBMO lie within the hyperellipsoid of the standard mutation, but the center of this hyperellipsoid is shifted by the vector $(b, \ldots, b)$. Again, $\boldsymbol{\xi}$ is mutated analog to the biased mutation of the standard BMO.

### 4.2.4  Comparison of Computational Effort of the Randomized Operators

Let $N$ be the number of strategy variables. Uncorrelated Gaussian mutation with one step size requires only $O(1)$ operations, uncorrelated Gaussian mutation with $N$ steps sizes $O(N)$ respectively. Correlated mutation needs $O(N^2)$ for the rotation of the mutation ellipsoid and is thus computationally expensive. The directed mutation requires $O(2N)$ operations, but the computation of the asymmetric probability density function is more complex. Also the BMO needs

$O(2N)$ operations, but is based on the simpler Gaussian mutation. In comparison to correlated mutation the BMO saves $\frac{N\cdot(N-1)}{2} - N$ additional strategy parameters. Chapter 4.5 reveals the power of the different approaches.

## 4.3   The Descent Direction Mutation Operator (DMO)

The BMO enables to bias mutations into beneficial directions. Under the assumption that the optimum lies on the extension of the direction between the center of consecutive generations, the idea arises to adjust the bias along this direction. More precisely, the descent direction mutation operator (DMO) adjusts the bias to the descent direction of the centers of two successive generations. Figure 4.5 illustrates the idea of the DMO. Let $\chi_t$ be the center of the population of generation $t$

$$\chi_t = \sum_{i=1}^{\mu} \mathbf{x}_i \tag{4.39}$$

The normalized descent direction $\boldsymbol{\xi}$ of two successive population centers $\chi_t$ and $\chi_{t+1}$ is

$$\boldsymbol{\xi} = \frac{\chi_{t+1} - \chi_t}{|\chi_{t+1} - \chi_t|} \tag{4.40}$$

Similar to the BMO, the DMO now becomes

$$\mathbf{z} := (\sigma_1 \mathcal{N}_1(0,1) + \xi_1\sigma_1, \ldots, \sigma_N \mathcal{N}_N(0,1) + \xi_N\sigma_N) \tag{4.41}$$

$$= (\sigma_1 \mathcal{N}_1(\xi_1, 1), \ldots, \sigma_N \mathcal{N}_N(\xi_N, 1)) \tag{4.42}$$

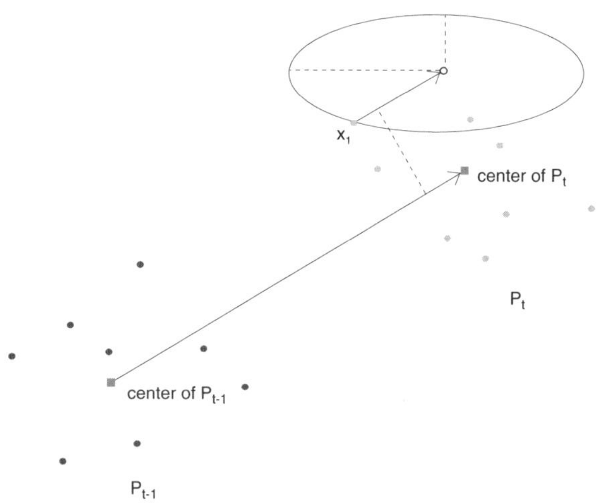


**Fig. 4.5.** Principle of the DMO in two dimensions: The vector reaching from the center of population $P_{t-1}$ to the center of the population $P_t$ defines the descent direction of the DMO, i.e. the shift of the center of the mutation ellipsoid

The DMO is reasonable as long as the assumption of locality is true: the estimated direction of the global optimum can be derived from the local information, i.e. the descent direction of two successive populations' centers. In the experimental analysis of section 4.5, we will examine the convergence properties of the DMO on practical problems.

## 4.4  Success Rate on Monotone Functions

One basic intuition behind biased mutation and in particular behind the DMO is that a good search direction in the recent past may be beneficial in the next generation. The precondition of a successful descent direction mutation is a monotone part of the function. It is probable that the BMO learns the descent direction, as the self-adaptive process will favor the latter. In the following we analyze the success rates and the behavior of a simplified DMO model on monotone functions. Let $f : \mathbb{R} \to \mathbb{R}$ be the monotone decreasing objective function to be minimized. A $(1+1)$-EA with the DMO can be modeled by the Markovian process $(X_t, \sigma_t)_{t \geq 0}$ generated by

$$X_{t+1} = \begin{cases} X_t + \sigma_t Z_t + \sigma_t b_t & \text{if } f(X_t + \sigma_t Z_t + \sigma_t b_t) < f(X_t) \\ X_t & \text{otherwise} \end{cases} \tag{4.43}$$

with $X_0 = x_0 \in \mathbb{R}$, $\sigma_0 > 0$, and

$$\sigma_{t+1} = \begin{cases} \gamma_1 \sigma_t & \text{if } f(X_t + \sigma_t Z_t + \sigma_t b_t) < f(X_t) \\ \gamma_2 \sigma_t & \text{otherwise} \end{cases} \tag{4.44}$$

and $\gamma_1 > 1$, $\gamma_2 \in ]0, 1[$, as well as

$$b_t = \frac{X_t - X_{t-1}}{|X_t - X_{t-1}|} \tag{4.45}$$

Each $Z_t, t \geq$ is independent and identically distributed with a marginal density function. We take the standard Gaussian distribution into account. For monotone decreasing functions on $\mathbb{R}$ it holds $f(x+k) \leq f(x)$ for a positive $k$. Consequently, for $\sigma_t b_t > 0$, i.e. self-adaptation has found the bias $b_t$ into the descent direction or the DMO approximates the latter, it holds

$$f(X_t + \sigma_t Z_t + \sigma_t b_t) \leq f(X_t + \sigma_t Z_t) \tag{4.46}$$

We can compare the success rate of biased mutation to the success rate of isotropic mutation.

**Theorem 4.1.** *Let $(p_s)_{ISO}$ and $(p_s)_{BMO}$ be the success rates of isotropic Gaussian mutation and the BMO and let $b_t$ be a bias into the descent direction. On a monotone function $f$, it holds $(p_s)_{ISO} < (p_s)_{BMO}$.*

*Proof.* For an individual $x$ on a monotone function $f$, improving mutations are only produced in the area $\mathcal{S} = \{y|y > x\}$. Let $Z(y)$ be the density function of the normal distribution, see equation 4.4 and $Z_b(y) = Z(y) + b$ is $Z(y)$ shifted by $b > 0$. The success rates are

$$(p_s)_{\text{ISO}} = P\{Z(y) \geq x\} = \int_x^\infty Z(y)dy \tag{4.47}$$

and

$$(p_s)_{\text{BMO}} = P\{Z_b(y) \geq x\} = P\{Z(y) \geq x - b_t\} = \int_{x-b_t}^\infty Z(y)dy \tag{4.48}$$

Since

$$\int_x^\infty Z(y)dy < \int_{x-b_t}^\infty Z(y)dy \tag{4.49}$$

we get $(p_s)_{\text{ISO}} < (p_s)_{\text{BMO}}$. $\qquad\Box$

This result holds for monotone decreasing functions $f : \mathbb{R} \to \mathbb{R}$. Furthermore, the increase of the success probability leads to a higher probability to increase the step sizes $(\sigma_{t+1} = \gamma_1 \sigma_t)$. Hence, the DMO is able to *move fast* along monotone parts of a function.

## 4.5   Experimental Analysis

In this chapter we analyze the BMO experimentally. As the theoretical analysis of evolutionary approaches is frequently constrained to simple conditions, i.e. very simple test functions and simple algorithms, the experimental analysis is able to give insight into capabilities and drawbacks under more complex conditions. Of course, those results cannot be generalized to every problem or every problem class, but they state useful hints about the relationship between problem class characteristics and algorithms. The following questions arise:

- is the BMO able to improve an ES on uni- and multimodal problems,
- is the BMO able to improve an ES on constrained problems,
- what are the differences between the BMO variants,
- what are the conditions and limitations of the self-adaptive parameter control of the BMO, e.g. the influence of population ratios and mutation parameters like $\gamma$.

In order to answer these questions, we conduct a sequence of experiments of self-adaptive ES variants on various uni-, multimodal and constrained problems.

We compare the performance of the BMO and the variants sBMO, cBMO, as well as the DMO. For the BMO we analyze the influence of the parameter $\gamma$. To get an impression of the influence of the population ratios and the selection pressure we test a sequence of different settings of population sizes $\mu$ and $\lambda$. We compare the performance of the BMO variants to the ES with uncorrelated Gaussian mutations and $N$ step sizes[2]. We do not compare our self-adaptive operators to the CMA-ES, since the latter is based on *derandomized* step size control.

### 4.5.1  Unconstrained Real Parameter Optimization

In this section we test the biased mutation on unconstrained real parameter optimization problems. The test suite consists of the following problems, each with $N = 10$ dimensions:

- sphere,
- double sum,
- rosenbrock,
- rastrigin,
- griewank,
- rosenbrock with noise.

The features of the functions are listed in the appendix, see A. The following experimental settings have been used. The initialization, termination conditions and the number of runs, i.e. 25 runs, are taken from the recommendations of the *CEC 2005 special session on real-parameter optimization* [149]. ES denotes the uncorrelated Gaussian mutation with $N$ step sizes.

| **Experimental settings** | |
|---|---|
| Population model | (15,100) |
| Mutation types | ES, BMO, sBMO, cBMO, DMO, $\gamma = 0.1$, $\tau_0 = \frac{1}{\sqrt{N}}, \tau_1 = \frac{1}{\sqrt{2\sqrt{N}}}$ |
| Crossover type | intermediate, $\rho = 2$ |
| Selection type | comma |
| Initialization | [-100,100], $\sigma^{(0)} = 0.1$ |
| Termination | 1000 generations, 2000 generations (rosenbrock) |
| Runs | 25 |

### Comparison of Variants

First of all, we test the BMO and its variants on a set of classical test functions. Table 4.1 shows the experimental results on the sphere function. The results of the experiments show that the BMO and the variants sBMO and cBMO do not achieve significant improvements. This also holds for the DMO. It turned out that the sBMO could not achieve remarkable results within the following experiments.

---

[2] This standard ES is abbreviated with *ES* in the tables and figures of this chapter.

**Table 4.1.** Experimental results of the BMO variants on the function sphere. The BMO achieves a better mean than the variants sBMO, cBMO and DMO. Its results are comparable to the results of the ES. But the unimodal monotone function does not require the bias. This is why the unbiased ES with only one step size (sES) achieves a much better accuracy.

|        | ES        | BMO         | sBMO        | cBMO        | DMO         |
|--------|-----------|-------------|-------------|-------------|-------------|
| best   | 2.36E-150 | 5.83E-149   | 2.12E-152   | 5.41E-145   | 1.80E-146   |
| median | 3.65E-146 | 3.82E-146   | 4.89E-145   | 2.75E-141   | 4.50E-144   |
| worst  | 3.60E-141 | 7.51E-143   | 9.22E-138   | 5.25E-138   | 5.89E-140   |
| mean   | 2.07E-142 | **5.33E-144** | 3.95E-139 | 2.42E-139   | 2.49E-141   |
| dev    | 1.07E-146 | 9.2699E-144 | 4.6107E-138 | 7.8575E-143 | 7.6016E-144 |

**Table 4.2.** Experimental results of the BMO variants on the function double sum. ES, cBMO and DMO show the same approximation quality.

|        | ES        | BMO        | cBMO      | DMO       |
|--------|-----------|------------|-----------|-----------|
| best   | 3.57E-47  | 1.61E-38   | 8.06E-40  | 5.15E-45  |
| median | 4.94E-39  | 1.71E-30   | 1.54E-36  | 6.59E-41  |
| worst  | 8.55E-36  | 1.53E-24   | 7.68E-34  | 1.58E-34  |
| mean   | 7.96E-37  | 1.02E-25   | 6.34E-35  | 6.39E-36  |
| dev    | 2.33E-36  | 3.588E-25  | 1.68E-34  | 3.16E-35  |

In table 4.2 the outcome of the experiments on the double sum function is presented. It turns out that the BMO achieved slightly worse results than the standard ES, cBMO and DMO. But the approximation of the optimum of the cBMO and the DMO are as good as the ES. As no considerable improvements on the unimodal functions could have been observed we now test the meta heuristics on multimodal functions.

First, we compare the behavior of the BMO variants on the rosenbrock function. Table 4.3 and figure 4.6 show the results we accomplished. It turns out that the BMO and the variant cBMO as well as the DMO deliver better results than the ES. All parameters *best, median, worst, mean* and *dev* are smaller for the

**Table 4.3.** Experimental results of the BMO variants on the function rosenbrock. BMO and its variants show much better results than the standard ES mutation.

|        | ES       | BMO       | sBMO     | cBMO      | DMO       |
|--------|----------|-----------|----------|-----------|-----------|
| best   | 0.00027  | 7.97E-10  | 0.02016  | 7.83E-11  | 1.79E-05  |
| median | 0.02734  | 1.22E-08  | 0.02435  | 3.45E-08  | 2.78E-05  |
| worst  | 0.19001  | 4.72E-08  | 0.98524  | 6.03E-05  | 0.00027   |
| mean   | 0.03412  | **1.47E-08** | 0,07384 | 2.48E-06 | 4.14E-05  |
| dev    | 0.03450  | 1.14E-08  | 0.19425  | 1.20E-05  | 5.15E-05  |

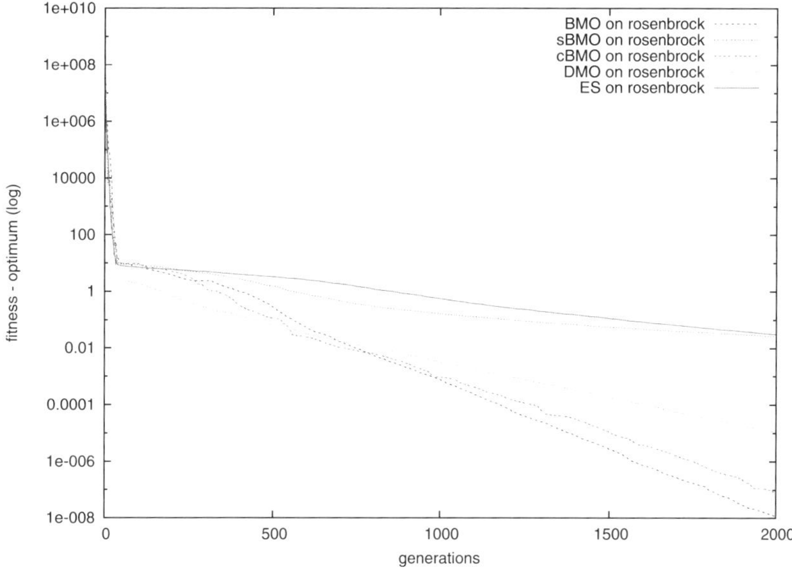


**Fig. 4.6.** Fitness development of typical runs of the ES, the BMO, its variants and the DMO on the rosenbrock function. BMO and cBMO show the fastest approximation capabilities while the standard ES is comparatively slow.

mentioned variants. The variant sBMO shows rather weak results, but is still better than the ES as the Wilcoxon rank-sum test will show in the following. These results lead to the conclusion that the additional degree of freedom obtained by the bias offers an advantage in the multimodal fitness landscape of the rosenbrock function.

## Wilcoxon Rank-Sum Test

To prove the significance of the experimental results, we perform a pairwise comparison of the operators with a Wilcoxon rank-sum test [68]. The Wilcoxon rank-sum test is non-parametric and tests the hypothesis that two sample sets are drawn from a single population, i.e. that their probability distributions are equal. The Wilcoxon rank-sum test does not require any assumptions like the student t-test. The latter is only applicable if the source distribution is a normal distribution, which is usually not the case for EAs. It is applicable if the distribution of the data is not known. The Wilcoxon rank-sum test is more sensitive for small data sets, as it is based on the combined ranking of both data sets. The summed rank of the smallest sample is the value for the level of significance calculation. The exact level of significance makes use of all possible permutations of ranks and can be approximated with a normal distribution.

|      | BMO       | sBMO   | cBMO      | DMO       |
|------|-----------|--------|-----------|-----------|
| ES   | 1.421E-09 | 0.1253 | 1.421E-09 | 1.603E-09 |
| BMO  |           |        | 1.421E-09 | 0.01301   | 1.421E-08 |
| sBMO |           |        |           | 1.421E-08 | 1.421E-09 |
| cBMO |           |        |           |           | 2.059E-08 |

We assume a significance level of $p_l = 0.05$. The above table shows the $p_w$-values, which indicate the probability that the null hypothesis, i.e. the two data sets show the same distribution, in particular the same median, holds. The comparison of the ES to the biased mutation variants reveals a significant superiority of the BMO ($1.4E - 9 << 0.05$), the cBMO ($1.4E - 9 << 0.05$) and the DMO ($1.6E - 9 << 0.05$). This is not the case for the sBMO. It exhibits the same distribution with probability $p \leq 0.1253$, which is not significant. Furthermore, the test shows that the BMO is significantly better than the sBMO, the DMO and even the cBMO ($0.01301 << 0.05$). The variants cBMO and DMO are significantly better than sBMO. BMO and cBMO cannot be distinguished with a significant probability. Consequently, we can assert the order 1. BMO and cBMO, 2. DMO, 3. sBMO, 4. ES.

The new strategy parameters have to be controlled adequately, otherwise the search will get stuck. The sBMO with an N-dimensional bias, but only one step size is not able to control the bias properly. We analyze population ratios and the mutation parameter $\gamma$ in the next paragraph. The comparison on the function rastrigin, see table 4.4, upper part, does not reveal any superiority of the biased mutation variants. The standard ES as well as the DMO only show average results during the approximation of the optimum. We observe a slight deterioration of the BMO and the DMO in comparison to the other variants. A similar behavior can be observed on the function griewank, see table 4.4, lower part. Neither an improvement nor a deterioration is obtained by the usage of the self-adaptive bias.

**Table 4.4.** Experimental results of the BMO variants on the function rastrigin (upper part) and the function griewank (lower part). No significant superiority of any variant can be observed.

|           | ES    | BMO   | cBMO  | DMO   |
|-----------|-------|-------|-------|-------|
| *rastrigin* |     |       |       |       |
| best      | 0.99  | 3.97  | 0.99  | 1.98  |
| median    | 2.98  | 11.93 | 2.98  | 5.96  |
| worst     | 6.96  | 45.76 | 6.96  | 11.93 |
| mean      | **3.22** | 16.55 | 3.58 | 6.76 |
| dev       | 1.80  | 10.20 | 1.64  | 2.79  |
| *griewank* |      |       |       |       |
| best      | 0     | 0     | 0     | 0     |
| median    | 0     | 0.009 | 0     | 0.014 |
| worst     | 0.022 | 0.036 | 0.017 | 0.145 |
| mean      | 0.004 | 0.013 | **0.003** | 0.023 |
| dev       | 0.006 | 0.013 | 0.006 | 0.029 |

**Table 4.5.** Parameter study of $\gamma$ for the BMO on the rosenbrock function. Settings around $\gamma = 0.1$ achieve the best approximation qualities.

|        | 0.001    | 0.01     | 0.05        | 0.1         | 0.15        | 0.5      | 1.0      |
|--------|----------|----------|-------------|-------------|-------------|----------|----------|
| best   | 1.54E-02 | 5.24E-03 | 3.95E-10    | 7.97E-10    | 6.53E-09    | 8.38E-05 | 8.80E-03 |
| median | 2.73E-02 | 1.87E-02 | 1.83E-08    | 1.22E-08    | 1.11E-07    | 1.59E-03 | 1.58E-02 |
| worst  | 9.16E-01 | 6.43E-01 | 5.69E-07    | 4.72E-08    | 2.89E-07    | 8.45E-03 | 8.41E-01 |
| mean   | 6.52E-02 | 4.59E-02 | **5.81E-08** | **1.47E-08** | **1.23E-07** | 1.89E-03 | 7.48E-02 |
| dev    | 1.77E-01 | 1.26E-01 | 1.19E-07    | 1.14E-08    | 7.65E-08    | 1.49E-03 | 2.04E-01 |

## Parameter Study of $\gamma$

Now we conduct a parameter study of the biased mutation parameter $\gamma$. Mutation parameters of strategy variables often have a strong impact on self-adaptation capabilities. Whether this is also the case for the BMO, table 4.5 answers. The parameter $\gamma$ controls the mutation strength of the bias coefficient vector $\boldsymbol{\xi}$. For $\gamma$ values between 0.001 and 1.0 have been tested. There is a strong influence of $\gamma$ on the quality of the results. In fact, only the values $\gamma \approx 0.1$, i.e. $\gamma = 0.05$, 0.1 and 0.15 let the BMO obtain a successful approximation of the optimum. For higher values the mutation is too strong and does not adapt the bias properly. For lower values the mutation is too weak and achieves no significant influence. As further experiments on other test functions reveal a similar optimal behavior for $\gamma$, we recommend settings around $\gamma \approx 0.1$ being aware that according to *no free lunch* other settings might be more appropriate for other problems.

## Selection Pressure and Population Ratios

Selection pressure and population ratios influence the population's diversity withing the strategy parameter search space. We vary the selection pressure and the population ratio in order to explore the BMO's self-adaptation capabilities. Table 4.6 shows the outcome of these experiments. As we can see, we obtain no improvement of the mean results with an increase of the selection pressure [(5,100);(15,300);(15,500)]. Among these three settings the (15,100) ratio obtains the results with the highest stability and lowest standard deviation. Only the (5,100)-ES achieves a better *best* and *median*. But as the distribution of the results is relatively high, it pays with a slightly higher standard deviation. An increase of offspring individuals is supposed to increase the population's diversity. But this has no positive impact on the achieved results. No improvements can be obtained with a high number of parents and a weak selection pressure. We explain these results with a weakened connection between strategy variables and fitness: a weak selection pressure allows solutions with comparatively bad strategy parameters to survive. Hence, an evolution of strategy parameters is almost impossible.

**Table 4.6.** Experimental population ratios analysis of the BMO on the rosenbrock function. The (15,100)-BMO achieves the best approximation quality. No improvement can be gathered with an increase of population sizes.

|        | (5,100) | (15,100) | (15,300) | (15,500) | (50,100) | (50,300) | (50,500) |
|--------|---------|----------|----------|----------|----------|----------|----------|
| best   | **6.61E-11** | 7.97E-10 | 9.55E-07 | 1.61E-05 | 0.00179 | 0.00169 | 0.00635 |
| median | **1.36E-09** | 1.22E-08 | 3.69E-06 | 0.00014 | 0.00352 | 0.00266 | 0.01373 |
| worst  | 2.97E-05 | **4.72E-08** | 0.00011 | 0.00088 | 4.02246 | 0.11635 | 1.53753 |
| mean   | 1.24E-06 | **1.47E-08** | 9.19E-06 | 0.00019 | 0.32561 | 0.00773 | 0.07479 |
| dev    | 5.93E-06 | 1.14E-08 | 2.30E-05 | 0.00019 | 1.10751 | 0.02271 | 0.30477 |

**Table 4.7.** Experimental results of the DMO with various population ratios on the rosenbrock function. The (5,100)- and the (15,100)-DMO show better approximation capabilities than the other variants.

|        | (5,100) | (15,100) | (15,300) | (15,500) | (50,100) | (50,300) | (50,500) |
|--------|---------|----------|----------|----------|----------|----------|----------|
| best   | 1.43E-05 | 1.79E-05 | 0.00050 | 0.00416 | 0.02784 | 0.03450 | 0.08350 |
| median | 3.20E-05 | 2.78E-05 | 0.00085 | 0.00634 | 0.04307 | 0.04294 | 0.09682 |
| worst  | 6.10E-05 | 0.00027 | 4.06074 | 0.92337 | 332.30653 | 103.10865 | 23.87899 |
| mean   | **3.49E-05** | **4.14E-05** | 0.45365 | 0.04305 | 14.26908 | 4.87227 | 1.86968 |
| dev    | 1.17E-05 | 5.15E-05 | 1.16190 | 0.18340 | 66.29393 | 20.68901 | 5.66927 |

We can summarize that a relatively high selection pressure can be recommended to achieve high quality results if many runs of the algorithm are possible. The probability for outstanding results among many runs is high. In this case the best solution can be saved. If a high reliability is necessary, we recommend the standard setting (15,100). Whether these results are also feasible for the DMO, the following table 4.7 tries to answer. The results roughly confirm the outcome of the previous experiments: the (5,100)- and the (15,100)-BMO achieve the best approximation of the optimum. But no outstanding best result can be obtained with the (5,100)-BMO. A decrease of the selection pressure or an increase of $\mu$ results in a deterioration as we can see from the experiments of the settings (15,300) to (15,500). On the other side an increase of $\lambda$ is an increase of the selection pressure and improves the results.

**Noisy Fitness Functions**

Noise in the fitness is a frequent condition of real-world optimization problems. Here, we test the biased mutation on the rosenbrock function with noise. The obtained results can be found in table 4.8. The experiments show that the biased mutation is robust against noise on the rosenbrock function. We observe a significant better behavior of the biased mutation variants in comparison to the standard ES, see the Wilcoxon rank-sum test below. Hence, the BMO can be recommended for fitness landscapes where noise is present. The same holds for the DMO. Although the mean of the cBMO is worse, the median is better

**Table 4.8.** Experimental results of the BMO variants on the function rosenbrock with noise. The BMO and the DMO are robust against noise.

|        | ES      | BMO   | cBMO    | DMO    |
|--------|---------|-------|---------|--------|
| best   | 0.449   | 0.008 | 0.223   | 1.004  |
| median | 5.236   | 0.027 | 2.864   | 2.876  |
| worst  | 172.983 | 7.693 | 593.256 | 81.778 |
| mean   | 20.890  | **0.378** | 42.305 | 7.298 |
| dev    | 45.245  | 1.535 | 137.020 | 16.511 |

than the median of the ES and the Wilcoxon rank-sum test reveals statistical superiority. The bad mean and worst values were produced by an outlier.

**Wilcoxon Rank-Sum Test**

|      | BMO       | cBMO      | DMO       |
|------|-----------|-----------|-----------|
| ES   | 1.841E-08 | 0.001462  | 0.000445  |
| BMO  |           | 4.006E-08 | 2.059E-08 |
| cBMO |           |           | 0.4728    |

In order to prove the statistical relevance of the test data, we again perform the Wilcoxon rank-sum test. BMO and DMO are better than the standard ES. Their median and mean are below the values of the ES and all $p_w$-values are smaller than $p_l = 0.05$. Statistical superiority also holds for the cBMO in spite of the worse mean value. But its median is better than the median of the ES. The worse mean was obviously produced by outliers. Not surprisingly, the BMO is significantly better than the cBMO and the DMO, while cBMO and DMO are not significantly distinguishable from each other.

We summarize the obtained *successful* results of biased mutation on the unconstrained real parameter optimization problems. Biased mutation achieves considerable improvements on the function rosenbrock. The same holds for the function rosenbrock with noise. No significant improvement or deterioration could be observed on the tested unimodal functions.

## 4.5.2  Climbing Ridges with Biased Mutation

In this section we test the biased mutation on two real parameter optimization problems with ridges. The ridge function class comprises functions of the following form:

$$f_{ridge}(\mathbf{x}) := x_1 - d \left( \sum_{i=2}^{N} x_i^2 \right)^{(\alpha/2)} \tag{4.50}$$

We assume that a bias along the ridges increases the speed of the optimization process and examine the following test functions ($d = 1$):

- sharp ridge, $\alpha = 1$,
- parabolic ridge, $\alpha = 2$.

The following experimental settings have been used.

| **Experimental settings** | |
| --- | --- |
| Population model | (15,100) |
| Mutation types | ES, BMO, cBMO, DMO, $\gamma = 0.1$, $\tau_0 = \frac{1}{\sqrt{N}}$, $\tau_1 = \frac{1}{\sqrt{2\sqrt{N}}}$ |
| Crossover type | intermediate, $\rho = 2$ |
| Selection type | comma |
| Initialization | [-100,100] |
| Termination | 1000 generations |
| Runs | 25 |

Earlier experiments showed [17] that self-adaptive step size control on ridge functions may result in premature step size reduction. These results could not be observed in our experiments.

## Comparison of Variants

The outcome of the BMO variants on the two ridge functions can be found in table 4.9. The experimental results confirm the above mentioned assumption: the bias increases the speed along the ridge. This holds for the parabolic ridge as well as for the sharp ridge. Furthermore, the experiments show that the BMO and cBMO are slightly faster than the DMO. Figure 4.7 shows the fitness development of typical runs of the variants on the parabolic ridge. The approximation speed of the biased variants BMO, cBMO and DMO is higher than the speed of the unbiased ES. The comparison does not reveal significant differences among the behavior of the successful biased variants.

**Table 4.9.** Experimental results of the BMO variants on the functions parabolic ridge (upper part) and sharp ridge (lower part). The BMO variants are significantly faster than standard Gaussian mutation.

| | ES | BMO | cBMO | DMO |
| --- | --- | --- | --- | --- |
| *parabolic* | | | | |
| best | -3.34E+190 | -5.69E+240 | -8.02E+239 | -2.13E+228 |
| median | -2.30E+187 | -4.16E+236 | -1.32E+236 | -7.98E+224 |
| worst | -7.54E+181 | -8.26E+231 | -8.90E+229 | -6.22E+219 |
| mean | -2.81E+189 | **-2.32E+239** | -5.37E+238 | -1.89E+227 |
| *sharp* | | | | |
| best | -3.52E+188 | -1.93E+234 | -5.46E+235 | -3.06E+226 |
| median | -2.40E+181 | -8.49E+229 | -2.79E+230 | -7.34E+221 |
| worst | -9.49E+174 | -6.22E+206 | -9.08E+210 | -6.24E+215 |
| mean | -2.52E+187 | -1.09E+233 | **-2.22E+234** | -1.44E+225 |

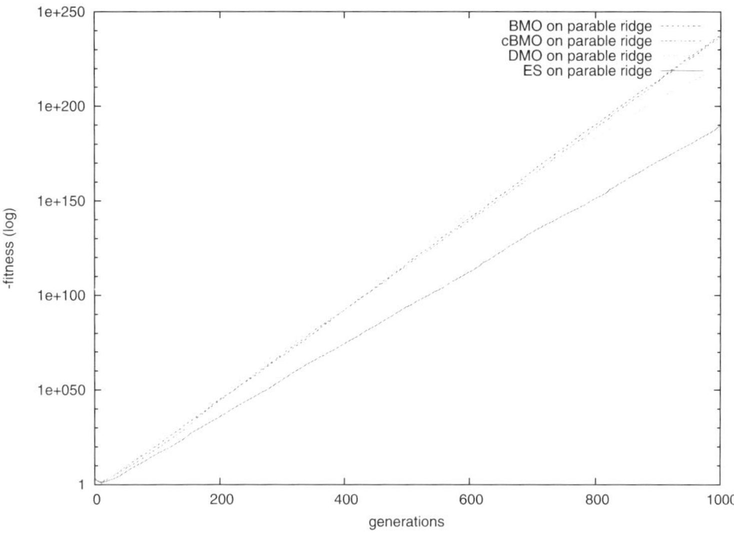

**Fig. 4.7.** Comparison of the biased mutation variants and the unbiased standard mutation on the parabolic ridge. The figure shows the absolute value of the fitness development of typical runs on a logarithmic scale and reveals the superiority of the variants BMO, cBMO and the DMO.

## Parameter Study of $\gamma$

Again, we test the influence of the strategy variable mutation parameter $\gamma$ on the function sharp ridge and observe remarkable effects, see table 4.10. Settings around the recommendation $\gamma = 0.1$, i.e. $0.05 \leq \gamma \leq 0.15$ achieve the best results. With these settings the BMO is much faster moving along the ridge axes in best, worst and average. In the range of $0.15 \leq \gamma \leq 0.05$ the probability to adjust the bias adequately is very high and results in an increase of the optimization speed. These settings for $\gamma$ correspond with the settings for other problem classes, i.e. the constrained problems, see section 4.5.3.

**Table 4.10.** Parameter study of $\gamma$ for the BMO on the function sharp ridge. Values around $\gamma \approx 0.1$ achieve the fastest convergence speed.

| | 0.001 | 0.01 | 0.05 | 0.1 | 0.15 | 0.5 | 1.0 |
|---|---|---|---|---|---|---|---|
| best | -1.7E+193 | -1.7E+193 | -4.0E+236 | -1.9E+234 | -7.4E+233 | -1.1E+216 | -1.9E+197 |
| median | -4.2E+181 | -4.2E+181 | -7.3E+227 | -8.4E+229 | -7.0E+226 | -1.6E+204 | -3.7E+190 |
| worst | -1.8E+167 | -1.8E+167 | -4.8E+215 | -6.2E+206 | -2.9E+216 | -9.5E+194 | -6.3E+180 |
| mean | -7.0E+191 | -7.0E+191 | **-3.3E+235** | **-1.0E+233** | **-3.0E+232** | -4.6E+214 | -8.0E+195 |

**Table 4.11.** Experimental population ratio analysis of the BMO on the function sharp ridge. A high selection pressure with a small number of parents like the (5,100)-BMO achieves the fastest approximation along the ridge.

| | (5,100) | (15,100) | (15,300) | (15,500) | (50,100) | (50,300) | (50,500) |
|---|---|---|---|---|---|---|---|
| best | -2.05E+304 | -1.93E+234 | -8.44E+107 | -1.56E+70 | -1.83E+122 | -1.28E+74 | -4.60E+51 |
| median | -8.47E+291 | -8.49E+229 | -2.17E+101 | -3.10E+63 | -4.26E+116 | -1.08E+71 | -8.99E+48 |
| worst | -1.07E+244 | -6.22E+206 | -1.72E+86 | -1.03E+54 | -3.16E+108 | -4.55E+66 | -4.00E+45 |
| mean | **-8.30E+302** | -1.09E+233 | -3.55E+106 | -7.18E+68 | -1.31E+121 | -1.31E+73 | -3.85E+50 |

## Selection Pressure and Population Ratios

The question of the influence of the population ratios answers table 4.11, where we test various population ratios of the BMO on the sharp ridge. It turns out that an increase of the selection pressure results in an increase of optimization speed. The (5,100)-ES achieves the best results. With this setting the optimization speed is the fastest.

A decrease of the number of offspring $\lambda$ keeping $\mu$ constant leads to an improvement. A decrease of $\mu$ keeping $\lambda$ constant also results in a better approximation quality. Consequently, diversity in the population does not seem to be the most important aspect optimizing the ridge functions, as the fastest approximation can be achieved with small population sizes.

### 4.5.3   Handling Constraints with Biased Mutation

For many optimization problems the search space is constrained due to a variety of practical conditions. For EAs with a self-adaptive step size mechanism like ES, it is not easy to find an optimum, which lies on the boundary of the feasible search space due to premature step size reduction. This results in premature fitness stagnation before approximating the optimum, also see [75] and chapter 7. For big step sizes the constrained search space cuts off a big area of success to reproduce better mutations than the parent. So the step sizes reduce self-adaptively before reaching the area of the optimum. Our experiments revealed the benefits of biased mutation for constraint handling. The feature of unbiasedness of mutation operators postulated by Beyer [16] is not advantageous for constrained search domains. The experiments reveal that the biased mutation, i.e. the BMO, the BMO variants and the DMO exhibit useful features, which prevent premature step size reduction while using the simple constraint handling technique *death penalty*[3]. Intuitively speaking, the reason for this behavior is the following: the distribution function is shifted along the constraint boundary so that the success area is not cut off.

---

[3] *Death penalty* discards infeasible solutions and produces new individuals until the whole offspring population is feasible.

| Experimental settings | |
|---|---|
| Population model | (15,100) |
| Mutation types | ES, BMO, cBMO, DMO, $\gamma = 0.1$, $\tau_0 = \frac{1}{\sqrt{N}}$, $\tau_1 = \frac{1}{\sqrt{2\sqrt{N}}}$ |
| Crossover type | intermediate, $\rho = 2$ |
| Selection type | comma |
| Initialization | [-100,100] |
| Constraint Handling | death penalty, BMO |
| Termination | 1000 generations (g04), 2000 generations (2.40) |
| Runs | 25 |

## Comparison of Variants

At first, we compare the algorithms on the constrained function 2.40, see table 4.12. It turns out that the BMO and the DMO are both able to approximate the optimum, while the standard ES with death penalty fails. The cBMO is worse than BMO, DMO and even *almost*[4] ES. The measured standard deviation of the DMO on 2.40 is smaller than the accuracy of our data structure allows. Hence, the DMO allows the best approximation of the optimum.

**Table 4.12.** Experimental results of the BMO variants on the constrained function 2.40. BMO and DMO are able to find the optimum in every run while the ES with death penalty fails.

| | ES | BMO | cBMO | DMO |
|---|---|---|---|---|
| best | -4985.69 | -5000.0 | -4982.02 | -5000.0 |
| median | -4948.04 | -5000.0 | -4883.70 | -5000.0 |
| worst | -4676.76 | -4999.99 | -4683.89 | -5000.0 |
| mean | -4897.72 | **-5000.0** | -4865.01 | **-5000.0** |
| dev | 96.69 | 0.0 | 91.99 | 0.0 |

Figure 4.8 compares the algorithms on problem 2.40 considering typical runs on a logarithmic scale. The standard ES with DP early suffers from premature step size reduction while BMO and DMO show logarithmically linear approximation of the optimum. Furthermore, the figure shows that the DMO shows a faster convergence than the BMO.

## Wilcoxon Rank-Sum Test

| | BMO | cBMO | DMO |
|---|---|---|---|
| vs. ES | 1.421E-09 | 0.09519 | 1.421E-09 |

The question about the significance of the experiments answers a Wilcoxon rank-sum test. It reveals that the BMO and the DMO are both better than the ES with DP (1.4E-09<<0.05), but the cBMO is not (0.09>0.05). Obviously, a *cube*

---

[4] See next Wilcoxon rank-sum test, $p_w = 0.09 > 0.05$.

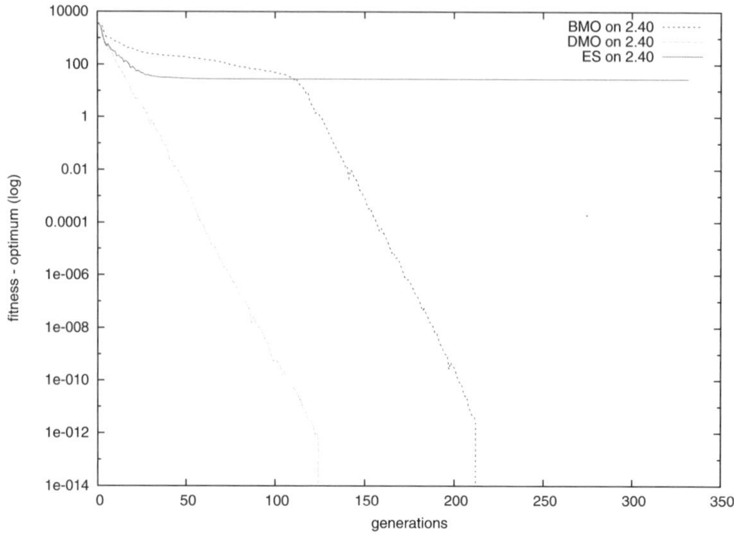

**Fig. 4.8.** Comparison between BMO, DMO and ES on problem 2.40. Typical runs are shown on a logarithmic scale. While the standard ES early suffers from premature steps size reduction, BMO and DMO show logarithmically linear approximation of the optimum.

**Table 4.13.** Experimental results of the BMO variants on the constrained function g04. BMO and DMO achieve a satisfying approximation while the ES is outperformed.

|        | ES           | BMO           | cBMO          | DMO           |
|--------|--------------|---------------|---------------|---------------|
| best   | -30665.53867 | -30665.53867  | -30665.53867  | -30665.53867  |
| median | -30665.53227 | -30665.53867  | -30665.53867  | -30665.53867  |
| worst  | -30644.23102 | -30665.53867  | -30662.83143  | -30665.53867  |
| mean   | -30664.23109 | **-30665.53867** | -30665.41105 | **-30665.53867** |
| dev    | 4.3651       | **0.0005**    | 0.5403        | **0.0003**    |

*bias* does not offer the necessary degree of freedom to shift the mutation ellipsoid appropriately.

A comparison of the BMO variants and ES with DP shows figure 4.9, where typical runs on problem g04 are drawn. BMO and DMO are able to approximate the optimum, while BMO is faster. This run of the ES fails. The results are similar to the behavior on problem 2.40. The results of 25 runs of all algorithms on problem g04 are shown in table 4.13. Again, the ES does not show satisfying results, the cBMO is slightly better. But BMO and DMO reach the optimum in every run with a small standard deviation. Similar results were obtained on other constrained functions.

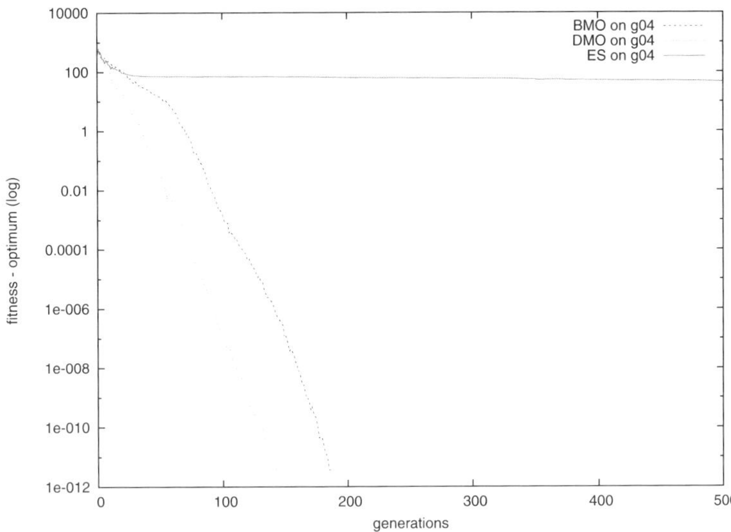

**Fig. 4.9.** Fitness development on function g04. Again, we compare typical runs of BMO, DMO and ES on a logarithmic scale. The approximation speed of DMO is better than the speed of BMO. ES does not approximate the optimum.

## Parameter Study of $\gamma$

We conduct an experimental study of the strategy variable mutation parameter $\gamma$ on problem g04, see table 4.14. A similar result like in the other $\gamma$ studies can be observed: parameter settings around $\gamma = 0.1$ like $\gamma = 0.05$ or $\gamma = 0.15$ show the best approximation behavior. Although many entries of the table are similar to each other, the standard deviation reveals the stability of the convergence properties.

**Table 4.14.** Parameter study of $\gamma$ for the BMO on the constrained function g04. Settings around $\gamma = 0.1$ maintain the fastest approximation capabilities.

|      | 0.0001 | 0.01 | 0.05 | 0.1 | 0.15 | 0.5 | 1.0 |
|------|--------|------|------|-----|------|-----|-----|
| best | -30665.538 | -30665.538 | -30665.538 | -30665.538 | -30665.538 | -30665.538 | -30665.538 |
| mean | -30658.772 | -30663.003 | **-30665.538** | **-30665.538** | **-30665.538** | -30665.538 | -30661.160 |
| dev  | 14.22802 | 6.72839 | **0.00056** | **0.00056** | **0.00039** | 0.00079 | 21.72168 |

## Selection Pressure and Population Ratios

We examine the behavior of various population sizes and ratios on the constrained function g04 and show the results in table 4.15. All results were satisfying with exception of the (50,100)-BMO. The ratio (50,100) is a very weak selection pressure. But a weak selection pressure slows down the approximation, so that

**Table 4.15.** Experimental results of the BMO with various populations ratios on the constrained function g04. Only the population ratio of (50,100) could not achieve good approximation results because of the weak selection pressure.

|      | (5,100) | (15,100) | (15,300) | (15,500) | (50,100) | (50,300) | (50,500) |
|------|---------|----------|----------|----------|----------|----------|----------|
| best | -30665.538 | -30665.538 | -30665.538 | -30665.538 | -30552.670 | -30665.538 | -30665.538 |
| mean | -30665.538 | -30665.538 | **-30665.538** | **-30665.538** | -30438.105 | **-30665.538** | **-30665.538** |
| dev  | 0.00069 | 0.00056 | **0.00039** | **0.00039** | 82.70956 | **0.00039** | **0.0** |

the quality is comparatively worse after 1000 generations. The similar behavior of the other settings does not allow a recommendation of a specific setting. But further tests on other functions showed that (15,100) and (15,300) are good choices.

## Interpretation of the BMO in Constrained Domains

The question arises, how the BMO is able to prevent the premature step size reduction in constrained search domains. Figure 4.10 shows the situation in the vicinity of the constraint boundary. As we will explain in chapter 7, the success probability $p_s$ for big step sizes $\sigma > d$ is small, $d$ is the distance of individual $x$ to the optimum. The mutation ellipsoid is cut off by the constraint boundary. The experiments show that the BMO produces mutations without decreasing the

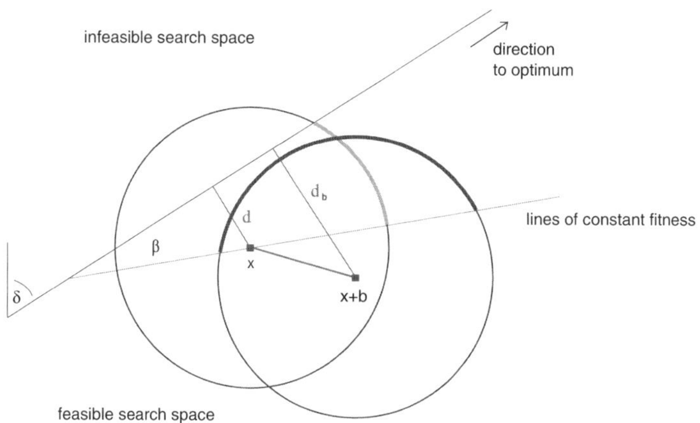


**Fig. 4.10.** The effect of the BMO in the vicinity of the infeasible search region: The bias is able to increase the success rate $p_s$. Mutations become possible that maintain a certain distance to the constraint boundary. A decrease of the step sizes is not advantageous anymore.

step sizes. Figure 4.10 tries to explain this behavior. We assume that mutations of $x$ and $x+b$ lie on the edges of the circles. The yellow circular arc is the success area of mutations without the bias. The bias $b$ can point into directions, which maintain a certain distance $d_b \geq d$ to the constraint boundary. The circle around $x + b$ represents BMO mutations. The red circular arc marks the success area of the BMO with shift $b$. This success area is much bigger and will be preferred in the self-adaptive process. Hence, the bias $b$ can adapt to directions, which increase the success rate $p_s$ without decreasing the step sizes.

## 4.6   Excursus: Self-Adaptive Mutation Operator Selection

The mutation operators are appropriate for different kinds of problems and search space characteristics. Which is the appropriate mutation operator for a given problem? We have seen various mutation operators for real-parameter optimization. In a series of experiments we tested the self-adaptive selection of the following mutation operators:

- isotropic mutation with $\sigma_i = 1$,
- uncorrelated mutation with $\sigma_i = N$,
- BMO,
- cBMO,
- DMO.

Each individual got an additional strategy parameter $mut_{type}$ determining the mutation operator to apply in the current generation. At the beginning $mut_{type}$ is randomly initialized. In the course of the evolutionary process this strategy parameter is mutated randomly with uniform distribution. The other strategy parameters like step size or bias were inherited as usual or randomly initialized if not available from particular operator changes.

We summarize the results of this experimental series. On the same test function no particular operator was preferred. Indeed, each possible operator was chosen randomly at the beginning, randomly changed during the first iterations and did not or very seldom change during the rest of the run. An explanation of this result is the following: Whenever a mutation operator with a useful set of strategy parameters (i.e. step size or bias) was randomly generated at the beginning, it created more successful offspring than an operator with random and in most cases worse initialization. Hence, the main result of these experiments is that the fast adaptation of strategy variables of random operators prevents potentially better operators from showing their capabilities. They were not able to adapt their strategy variables. We think it would be possible to overcome this problem by evolving all operators simultaneously for short time periods under similar conditions and select the mutation operators after this evaluation period. But this modification is supposed to be inefficient because of the additional fitness evolutions during simultaneous periods.

## 4.7  Summary

1. Various mutation operators for ES exist, from the randomized standard Gaussian mutation of ES to the derandomized CMA.
2. In order to minimize computational and implementation effort while offering a greater flexibility than unbiased mutation, the BMO was developed. Unlike the principle of unbiasedness of mutation the BMO was introduced as a self-adaptive operator for biasing mutations into beneficial directions.
3. The operator makes use of a self-adaptive bias coefficient vector $\xi$, which determines the direction of the bias. The absolute bias is computed by multiplication with the step sizes. A number of variants like the sBMO and cBMO have been introduced.
4. Another variant is the DMO whose bias is not controlled self-adaptively, but with the help of the descent direction estimated by two successive populations' centers. This bias adaptation technique is reasonable as long as the assumption of locality is true, i.e. the estimated direction to the global optimum can be derived from the local information.
5. Under the assumption that the bias direction can be determined self-adaptively or with the help of the DMO technique, the intuition that biased mutation is superior to unbiased mutation on monotone functions or on monotone parts of functions could be confirmed theoretically.
6. Experimental analysis on typical test functions showed the superiority of biased mutations on multimodal functions. On rosenbrock and rosenbrock with noise the BMO variants and the DMO achieve considerable improvements. The biased techniques also allow to speed up the evolution along the ridge of the functions sharp and parabolic ridge.
7. On constrained functions the BMO and the DMO improve the convergence to the optimum, while standard methods like DP suffer from premature step size reduction.
8. In all experiments the population ratio of (15,100) shows a high stability of the BMO and DMO results.
9. The mutation of the bias coefficient $\xi$ with settings around $\gamma \approx 0.1$ achieves the best experimental results.
10. Statistical tests, i.e. the Wilcoxon rank-sum test on selected experimental data, validated the relevance of the experimental results.

# 5 Self-Adaptive Inversion Mutation

Inversion mutation is a mutation operator for combinatorial search domains. It is based on the random change of edges. A famous example for a combinatorial problem is the traveling salesman problem (TSP). Inversion mutation is a typical genetic operator for combinatorial problems like the TSP. It reverses the tour between two randomly chosen cities. Here, we propose a self-adaptive variant of inversion mutation. Self-adaptation is a successful control technique for the mutation strength, see chapter 3. Up to now, the concept of self-adaptation has not been applied to inversion mutation. As the number of successive inversion mutation operator applications can be seen as the mutation strength, we propose to control this parameter self-adaptively. In this chapter we introduce a self-adaptive variant of inversion mutation (SA-INV). We prove the convergence of inversion mutation and its self-adaptive variant and show experimentally that SA-INV speeds up the evolutionary process, in particular at the beginning of the search. Later we encounter the *strategy bound problem* and introduce a heuristic to overcome it.

## 5.1 Introduction

At first, we introduce the TSP and give an overview of evolutionary combinatorial optimization and self-adaptation for discrete strategy variables.

### 5.1.1 The Traveling Salesman Problem

The TSP is the problem to minimize the length of a salesman's tour visiting every city of a given set only once and then returning back to the point where he started. It is defined as follows.

**Definition 5.1 (Traveling Salesman Problem)**
*Let $C$ be a set of $N$ cities with distances $d(c_i, c_j) \in \mathbb{R}$ for each pair of cities $c_i, c_j \in C$. An optimal solution of the TSP is the shortest tour $\pi^*$ of $C$, i.e. a permutation $\pi : [1, \dots, N] \longmapsto [1, \dots, N]$ with minimum length $l = \sum_{i=1}^{N-1} d\left(c_{\pi(i)}, c_{\pi(i+1)}\right) + d\left(c_{\pi(N)}, c_{\pi(1)}\right)$.*

O. Kramer: Self-Adaptive Heuristics for Evolutionary Computation, SCI 147, pp. 81–95, 2008.
springerlink.com

TSP if a very famous combinatorial optimization problem with dozens of solution approaches. The exact solution of bigger TSP instances is impractical since the problem is NP-complete [45]. With dynamic programming a solution can be computed in time $O(n^2 \cdot 2^n)$, which is still impractical. Other exact methods to solve the TSP are branch-and-bound and linear programming approaches. Various heuristics exist to solve the TSP reaching from nearest-neighbor strategies to the 2-opt heuristic. 2-opt is a famous heuristic replacing two longer random edges by two shorter ones [26]. The concept is based on the idea of inversion mutation: we select a subtour, reverse it and accept an improvement. An approach similar to simulated annealing accepts improvements with a certain probability, e.g. recently theoretically analyzed by Meer [89]. He constructs a TSP instance for which simulated annealing outperforms a metropolis algorithm with a fixed temperature using the 2-opt heuristic.

### 5.1.2  Evolutionary Combinatorial Optimization

Combinatorial optimization problems are believed not to be efficiently solvable, most problems are NP-hard[1]. Examples for combinatorial problems are the knapsack problem or the TSP. For the latter the solution of trying all permutations is impractical as the number of possible solutions is $n!$. In the traditional representation the queue of integers represents the order of the tour in which the cities have to be visited. To avoid multiple nodes, which would happen after the application of n-point crossover, partially mapped crossover (PMX) is used, see chapter 6. The random keys approach avoids infeasible solutions [10], [146]: Each city is assigned with a random number $x \in [0, 1)$. The chromosome is produced by visiting the nodes in ascending order. As this order is always well-defined as long as no equal numbers exist, crossover operators like N-point crossover can be applied. Other evolutionary approaches are successful like memetic algorithms, which combine EAs with local optimization methods, e.g. 2-opt.

### 5.1.3  Self-Adaptation for Discrete Strategy Variables

Not many EAs with discrete strategy variables exist. In chapter 6 we introduce self-adaptive crossover. Self-adaptive crossover controls the position of crossover points which are represented by integers [74]. Punctuated crossover by Schaffer and Morishima [128] makes use of a bit string of discrete strategy variables which represent location and number of crossover points. In section 4.6 we examined the self-adaptive selection of mutation operators for ES. Spears [147] introduced a similar approach for the selection of crossover operators. Bäck [7], Smith [142], Fogarty [144] and Stone [148] introduced self-adaptive approaches for mutation rates.

---

[1] NP-hard (nondeterministic polynomial-time hard): a problem H is NP-hard iff there is an NP-complete problem L that is polynomial time Turing-reducible to H, i.e. $L \leq_T H$.

## 5.2  Self-Adaptive Inversion Mutation

The *inversion mutation* is a mutation operator for combinatorial problems. It performs a mutual change of the connections of two entity pairs. Taking the TSP into account *inversion mutation* means that the part between the cities $p_1 - 1$ and $p_2 + 1$ is passed through reversely. Cities $p_1 - 1$ and $p_2$ as well as $p_1$ and $p_2 + 1$ are connected to each other after the procedure. INV is defined as follows.

**Definition 5.2 (Inversion Mutation, INV)**
*Let $p_1, p_2 \in \mathbb{N}_{>0}$ be two randomly chosen points with $1 \leq p_1 \leq p_2 \leq N$ and let $\pi$ be a permutation of cities $\pi = (c_1, \ldots, c_{p_1}, c_{p_1+1}, \ldots, c_{p_2-1}, c_{p_2}, \ldots, c_N)$. Inversion mutation inverses the part between $p_1$ and $p_2$ such that*

$$\pi' = INV(\pi) = (c_1, \ldots, c_{p_2}, c_{p_2-1}, \ldots, c_{p_1+1}, c_{p_1}, \ldots, c_N) \tag{5.1}$$

Figure 5.1 shows a simple example of a successful application of INV. The swap of the edges $\overline{AC}$ and $\overline{BD}$ to $\overline{AB}$ and $\overline{CD}$ results in a shorter tour.

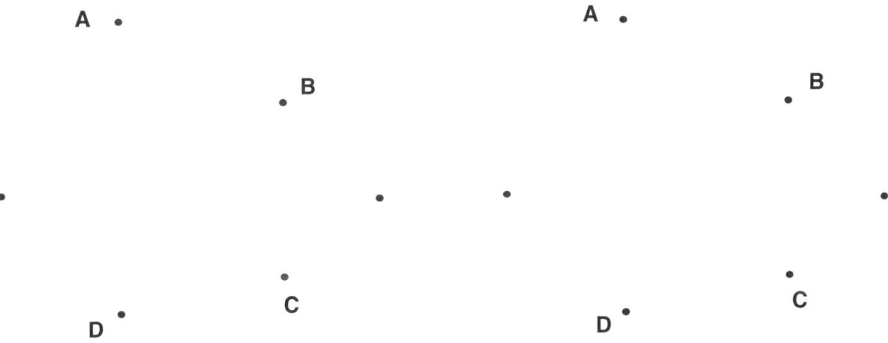


**Fig. 5.1.** Inversion mutation changes the connections between two pairs of nodes A, B, C and D. In this example, the swap of edges results in a shorter tour.

The number $k$ of successive applications of the inversion mutation operator can be seen as the mutation strength. $INV_k$ is the notation for $k$ successive INV-operations. The idea of SA-INV is to evolve $k$ self-adaptively. This leads to the following formalization of SA-INV.

**Definition 5.3 (Self-Adaptive Inversion Mutation, SA-INV)**
*Let $k_{t-1}$ be the number of successive INV-operations of the last generation. Then $k_t$ is mutated $k_t = mut(k_{t-1})$ and SA-INV is defined as*

$$\pi' = SA\text{-}INV(\pi) = INV_{k_t}(\ldots INV_1(\pi)). \tag{5.2}$$

Intuitively, as for nearly every optimization problem bigger mutations seem to be advantageous at the beginning of the search in order to explore the search domain. Later smaller steps make sense in order to optimize locally and to avoid the destruction of close-to-optimum solutions. Hence, it seems useful to adapt the number $k$ of successive applications of inversion mutation. For the strategy variable mutation of $k$ we use meta-EP mutation, see section 4.1.8,

$$k' := k + \gamma \cdot \mathcal{N}(0, 1). \tag{5.3}$$

Figure 5.2 shows the pseudocode of SA-INV.

| | |
|---|---|
| 1 | Mutate $k$ |
| 2 | **For** i=0 **to** $k$ |
| 3 | Randomly choose points $p_1$ and $p_2$ |
| 4 | Inverse part between $p_1$ and $p_2$ of tour $\pi$ |
| 5 | **Next** |

**Fig. 5.2.** Pseudocode of the SA-INV operator. The strategy variable $k$ determines the number of edge swaps. An edge swap is the inversion of the part between the randomly chosen pairs, i.e. edges $(p_1 - 1, p_1)$ and $(p_2, p_2 + 1)$.


## 5.3   Convergence Properties

In the following we prove that an EA using the INV-Operator finds the global optimum in finite time. As stated by Rudolph [125] and Eiben [36] it is not necessary to build an quantitatively exact Markov model for each EA variant. Instead, we can use a qualitative model. Let $\mathcal{X}$ be the search space and $(x_1, \ldots, x_n) \in \mathcal{X}^n$ denote the population of parents. Parent selection is abbreviated with *mat*, recombination with *reco*, mutation with *mut* and selection with *sel*. Our convergence proof is based on theorem 5.4. So, we have to ensure the following assumptions $(A_1)$ to $(A_4)$ [125]:

- $(A_1)$ $\forall x \in (x_1, \ldots, x_n) : P\{x \in reco(mat(x_1, \ldots, x_n))\} \geq \delta_r > 0$.
- $(A_2)$ For every pair $x, y \in \mathcal{X}$ there exists a finite path $x_1, x_2, \ldots, x_l$ of pairwise distinct points with $x_1 = x$ and $x_l = y$ such that $P\{x_{i+1} = mut(x_i)\} \geq \delta_m > 0$ for all $i = 1, \ldots, l - 1$.
- $(A_2')$ For every pair $x, y \in \mathcal{X}$ holds $P\{y = mut(x)\} \geq \delta_m > 0$.
- $(A_3)$ $\forall x \in (x_1, \ldots, x_l) : P\{x \in sel(x_1, \ldots, x_l)\} \geq \delta_s > 0$.
- $(A_4)$ Let $v_l^*(x_1, \ldots, x_l) = \max\{f(x_i) : i = 1, \ldots, l\}$ denote the best fitness value within a population of $l$ individuals ($l \geq n$). The selection method fulfills the condition

$$P\{v_n^*(sel(x_1, \ldots, x_l)) = v_l^*(x_1, \ldots, x_l)\} = 1. \tag{5.4}$$

**Theorem 5.4** *(Rudolph [125]). If the assumptions $(A_1)$, $(A_2)$ and $(A_3)$ are valid, then the EA visits the global optimum after a finite number of iterations with probability one, regardless of the initialization. If assumption $(A_4)$ is valid additionally and the selection method chooses from parents as well as offspring, then the EA converges completely and in mean to the global optimum regardless of the initialization.*

For the definitions of the terms *converges completely* and *converges in mean*, see section 2.1.2. We prove the global convergence of the following EA using INV as mutation operator:

1. recombination: no recombination is used, i.e. individuals for mutation are selected with uniform distribution,
2. mutation: INV/SA-INV see definitions 5.2 and 5.3,
3. selection: comma-selection with modification, i.e. comma-selection selects the $\mu - 1$ best parents from the $\lambda$ descendants and one parent randomly with uniform distribution and probability $p^* \geq 1/\lambda$.

**Theorem 5.5** *The EA with inversion mutation (INV) as defined above converges with probability 1 to the global optimum after a finite number of iterations.*

*Proof.* We make use of theorem 5.4 and have to ensure that assumptions $(A_1)$ to $(A_3)$ are valid. Assumption $(A_1)$ is valid, because $mat(x_1, \ldots, x_n) = x_i$, $0 \leq i \leq n$ with $P(X = i) = 1/n > 0$. Since no recombination is used $reco(x_i) = x_i$ with probability 1. So, $\forall x \in (x_1, \ldots, x_n)$ :

$$P\{x \in reco(mat(x_1, \ldots, x_n))\} = 1/n = \delta_r > 0 \qquad (5.5)$$

To prove assumption $(A_2)$ we have to show that for every tour $\pi_x$ and tour $\pi_y$: $\exists$ a finite sequence $INV_1, \ldots, INV_r$ of INV-operations such that $INV_r(\ldots INV_1 (\pi_x)) = \pi_y$. We use the following lemma 5.6.

**Lemma 5.6** *Let $\pi$ be a tour with cities $c_q$ with rank $x$ and $c_s$ with rank $y$, $x > y$. $\forall$ $c_q, c_s$ $\exists$ a finite sequence $INV_1, \ldots, INV_r$ of INV-operations with $\pi' = INV_r (\ldots INV_1(\pi))$ such that $\pi'(q) = y$, with probability $\delta_{m'(INV)} > 0$.*

*Proof.* It holds $x = \pi(q)$ and $y = \pi(s)$. For $x > y$, a sequence of $r = x - y$ successive $INV(\pi, i, j)$-operations[2] on neighbored cities moves city $c_q$ from rank $x$ to rank $y$, i.e.

$$INV(\pi, x, x - 1), INV(\pi_2, x - 1, x - 2), \ldots, INV(\pi_r, y + 1, y) \qquad (5.6)$$

Since parameters $i$ and $j$ of the INV-operator are chosen randomly with uniform distribution, the probability to choose them is $P\{i = x, j = x - 1\} = 1/(N(N - 1))$. Hence, the probability of the successive INV-operations is

---

[2] $INV(\pi, i, j)$ is inversion mutation of tour $\pi$ with *inversion points* $p_1 = i$ and $p_2 = j$.

$$\delta_{m'(\text{INV})} = P\{i_1 = x, j_1 = x - 1, \ldots, i_r = y + 1, j_r = y\} = \left(\frac{1}{N(N-1)}\right)^r$$

$$(5.7)$$

with the difference of ranks $r = x - y$. The order of the other cities is not changed by the neighbored INV-operations. □

If a sequence of mutations can move every city $c_q$ with rank $x$ to every rank $y$ as proven with lemma 5.6, there must exist a sequence of mutations that transfers tour $\pi_x$ into $\pi_y$ with probability

$$\delta_{m(\text{INV})} = P\{\pi_x \to \pi_y\} \geq \left(\frac{1}{N(N-1)}\right)^{r_{max}\cdot(N-1)}$$

$$(5.8)$$

with $r_{max} = \max\{r_i | x_i - y_i, 1 \leq i \leq N\} \leq N - 1$ and in the worst case $N - 1$ cities have to be moved by at most $r_{max}$. Hence, the probability for a finite path of pairwise distinct permutations $P\{\pi_{x_{i+1}} = mut(\pi_{x_i})\} \geq \delta_{m(\text{INV})} > 0$ and assumption (A$_2$) is valid. Assumption (A$_3$) is valid with $\delta_s = 1/\lambda > 0$. We had to modify the comma selection in this kind of way to ensure convergence and are convinced that the modification will not influence our experimental results.[3] Hence, the EA using the INV operator converges with probability 1 to the global optimum after a finite number of iterations. Assumption (A$_4$) would be valid for a 1-elitist strategy, i.e. the best individual is always kept in the population. In particular it is valid for plus selection and guarantees that the EA converges completely and in mean to the global optimum after a finite number of iterations. But this is not the case for comma selection. □

**Theorem 5.7** *The EA with self-adaptive inversion mutation (SA-INV) converges with probability 1 to the optimum after a finite number of iterations.*

*Proof.* Assumptions (A$_1$) and (A$_3$) are valid analog to the proof of theorem 5.5. Let $P_{k=1}$ be the probability for $k = 1$. As SA-INV uses Gaussian mutation with rounding procedures to mutate strategy parameter $k$, it holds $P\{k = 1\} > 0$ and according to the argumentation of theorem 5.5's proof with $k = 1$,

$$\delta_{m(\text{SA-INV})} = P\{k = 1\}^{r_{max}(N-1)} \cdot \delta_{m(\text{INV})} > 0$$

$$(5.9)$$

Hence, it holds $P\{\pi_{x_{i+1}} = mut(\pi_{x_i})\} \geq \delta_{m(\text{SA-INV})} > 0$ and (A$_2$) is valid. □

## 5.4   Experimental Analysis

In this section we analyze the SA-INV operator empirically. The analysis contains a comparison with the standard inversion mutation (INV) with constant number of edge swaps in each generation. We conduct the experiments on three

---

[3] Assumption (A$_3$) is also valid for fitness proportional and tournament selection.

TSP instances of the TSPlib of Reinelt [117], *berlin52, bier127* and *gr666*. Furthermore, we analyze the development of the strategy parameter $k$. For our experiments we make use of the following settings.

| Experimental settings | |
| --- | --- |
| Population model | (15,100) |
| Mutation type | SA-INV, INV, $\gamma = 1, 2, 5$ |
| Crossover type | no crossover |
| Selection type | comma |
| Initialization | random, $\nu = 5, 10, 20$ |
| Termination | 1 000 generations |
| Runs | 25 |

In order to avoid side effects we do not apply crossover. We use the following notation: SA-INV$_{\nu-\gamma}$ is the self-adaptive inversion mutation with the initial setting $k^{(0)} = \nu$ and the strategy mutation strength $\gamma$ and INV$_\nu$ is the common inversion mutation with constant $k = \nu$ edge swaps.

### 5.4.1   TSP Instance *Berlin52*

At first, we analyze SA-INV on the 2-dimensional *Euclidean distance* problem *berlin52*. Table 5.1 shows the results after 10 generations in order to show the behavior at the beginning of the search. The table clearly shows that SA-INV$_{5-1}$ achieves the best results after 10 generations. Within this time horizon INV$_1$ is slower with only one edge swap for a solution each generation. INV$_{10}$ is as faster than INV$_1$, because 10 swaps each generation produce better results, at least within this small timespan. The situation will change rapidly a couple of generations later, see figure 5.3. The lower part of the figure shows the fitness development of the SA-INV and INV variants. SA-INV$_{5-1}$ is the fastest variant at the beginning until INV$_1$ *overtakes* its approximation. We call the reason for that *strategy bound problem* and explain it in section 5.5.

While INV$_{10}$ is comparatively fast at the beginning, the 10 swaps destroy the solutions in later generations and hinder any further approximation of the optimum. This situation is similar to the comparison between an ES with constant

**Table 5.1.** Experimental analysis of SA-INV on TSP instance *berlin52* with 52 *cities*. SA-INV$_{5-1}$ is faster than INV$_1$ after 10 generations. The other SA-INV variants began with a too high number of edge swaps.

| | SA-INV$_{5-1}$ | SA-INV$_{10-2}$ | SA-INV$_{20-5}$ | INV$_1$ | INV$_{10}$ |
| --- | --- | --- | --- | --- | --- |
| best | 17027.32 | 17677.79 | 17493.31 | 16870.67 | 18136.78 |
| median | 18819.53 | 19123.82 | 19760.08 | 19526.80 | 19962.73 |
| worst | 20581.50 | 22221.77 | 25793.67 | 25586.68 | 21249.91 |
| mean | **18751.74** | 19365.57 | 20429.49 | 20111.88 | 19809.41 |
| dev | 826.00 | 1365.08 | 2557.71 | 2014.32 | 807.20 |

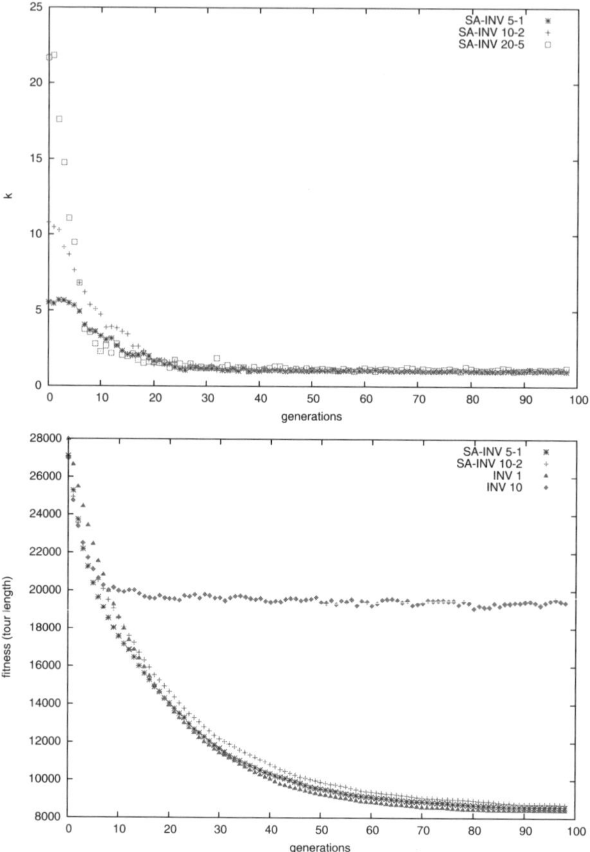

**Fig. 5.3.** SA-INV on problem *berlin52*. Upper part: Development of SA-INV strategy parameter $k$, averaged over 25 runs. Starting from different values for $k$, the runs show a similar development after 20 generations. Lower part: Fitness development of SA-INV and INV averaged over 25 runs. SA-INV$_{5-1}$ is the fastest variant in the first generations until INV$_1$ overtakes its approximation.

mutation strength and with self-adaptation, see figure 3.5. In the first 10 generations INV$_{10}$ develops similarly to SA-INV$_{10-2}$. But then SA-INV$_{10-2}$ reduces $k < 10$. Figure 5.3, upper part, shows the development of $k$. The self-adaptive process is able to control the *mutation strength* $k$ independently of the initial values $\nu = 5, 10, 20$: after generation 20 all developments proceed similarly to each other. Parameter $k$ almost constantly decreases to 1.

### 5.4.2   TSP Instance *Bier127*

Problem *bier127* is an *Euclidean distance* problem of 127 beer gardens in Augsburg. Table 5.2 presents the comparison of the SA-INV and INV variants after

**Table 5.2.** Experimental analysis of SA-INV on TSP instance *bier127* with 127 *cities* after 25 generations. All self-adaptive variants are faster than standard INV with fixed $k$.

|        | SA-INV$_{5-1}$ | SA-INV$_{10-2}$ | SA-INV$_{20-5}$ | INV$_1$ | INV$_{10}$ |
|--------|------------|------------|------------|-----------|-----------|
| best   | 353797.61  | 365157.31  | 372974.76  | 394652.11 | 403258.10 |
| median | 393111.29  | 393376.07  | 391979.44  | 414801.92 | 424457.13 |
| worst  | 410781.74  | 478352.23  | 427454.29  | 442159.83 | 438986.51 |
| mean   | **387801.57** | 397354.46 | 392307.67 | 416232.81 | 424336.54 |
| dev    | 15583.604  | 23499.15   | 14342.49   | 13123.65  | 8876.50   |

25 generations. All variants of the SA-INV are better than the standard INV operator concerning the *mean* and the *best* results. SA-INV$_{5-1}$ is the fastest variant, see also figure 5.4. Between generations 70 to 80, INV$_1$ reaches the same approximation quality, again we hold the *strategy bound problem*, see section 5.5, responsible for this. The *median* of the self-adaptive variants is approximately the same. The results of standard INV with $\nu = 10$ are poor. The number of ten edge exchanges seems to be too high in order to save the structure of the solutions, i.e. not to destroy them. This behavior is the basis for the success of SA-INV, which is able to reduce the mutation strength appropriately.

Figure 5.4 shows the development of the strategy parameter $k$ during the first 100 generations and the fitness development in the same time slot. For all SA-INV variants $k$ increases within the first generations, very demonstrative in the case of SA-INV$_{5-1}$. This shows that the self-adaptation process is working properly: an increase of the mutation strength ($k > 5$) is desirable at the beginning of the search. Later the mutation strength developments of all SA-INV variants show similar properties. SA-INV$_{20-5}$ decreases its $k$ faster between generation $t \approx 10$ and $t \approx 30$. We hold the higher strategy mutation parameter $\gamma$ responsible for that. Similar to instance *berlin52*, INV$_{10}$ suffers from fitness stagnation, here around $t \approx 20$. SA-INV$_{5-1}$ is better than INV$_1$ until generation $t \approx 80$.

### 5.4.3   TSP Instance *Gr666*

Problem *gr666*, also from TSPlib, is the largest TSP instance considered in this work. It is a *geographical distance* problem of 666 cities spread over the world. The experimental results after 70 generations can be found in table 5.3.

Also here, the picture drawn from the previous experiments is confirmed. The self-adaptive variants perform better than the standard ones, concerning the *best*, the *median*, the *worst* and the average results. We do not observe any significant differences between the three tested variants. INV$_{10}$ produces the worst results as the big mutation strength destroys solutions in later phases of the search.

In figure 5.5 the average development of strategy parameter $k$ and the fitness development over 25 runs and 100 generations of SA-INV and INV are shown. The smoothness of the development of $k$ over successive generations is worse than the smoothness on the previous problems. The reason is probably the size

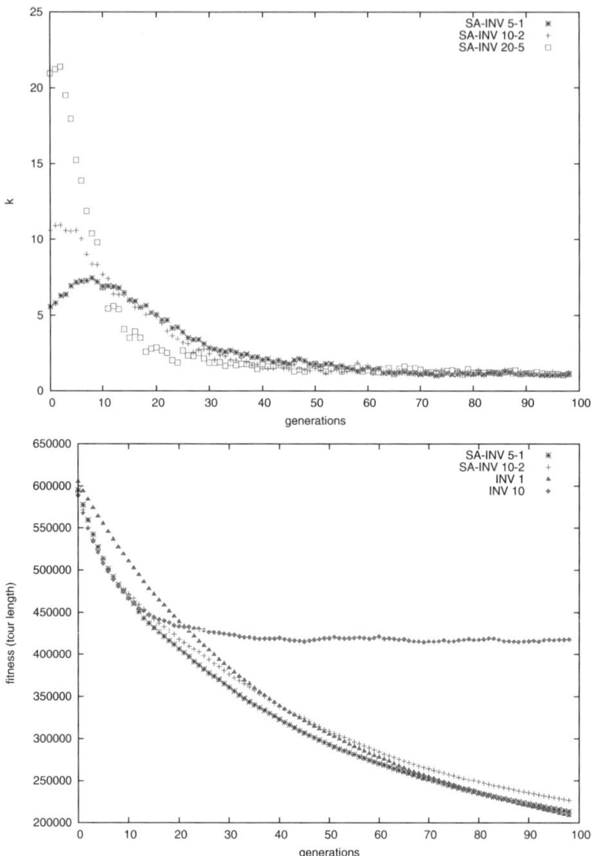

**Fig. 5.4.** SA-INV on problem *bier127*. Upper part: Development of $k$ averaged over 25 runs. For all SA-INV variants the *mutation strength* increases in the first generations. Lower part: Fitness development of SA-INV and INV averaged over 25 runs. $\text{SA-INV}_{5-1}$ is the fastest variant until generation $t \approx 80$. $\text{INV}_{10}$ is very fast at the beginning, but suffers from fitness stagnation after generation $t \approx 20$.

**Table 5.3.** Experimental analysis of SA-INV on problem *gr666* with 666 *cities*. All SA-INV variants are faster than the standard INV operators with fixed $k$.

|        | $\text{SA-INV}_{5-1}$ | $\text{SA-INV}_{10-2}$ | $\text{SA-INV}_{20-5}$ | $\text{INV}_1$ | $\text{INV}_{10}$ |
|--------|------------|------------|------------|------------|------------|
| best   | 3074023    | 3062252    | 3023938    | 3198055    | 3270915    |
| median | 3150807    | 3137811    | 3189521    | 3340901    | 3353568    |
| worst  | 3382585    | 3497183    | 3281412    | 3521550    | 3458274    |
| mean   | 3180473.08 | **3159244.16** | 3180724.16 | 3347929.96 | 3355142.4 |
| dev    | 84830.42   | 87259.16   | 73612.26   | 86277.86   | 47431.11   |

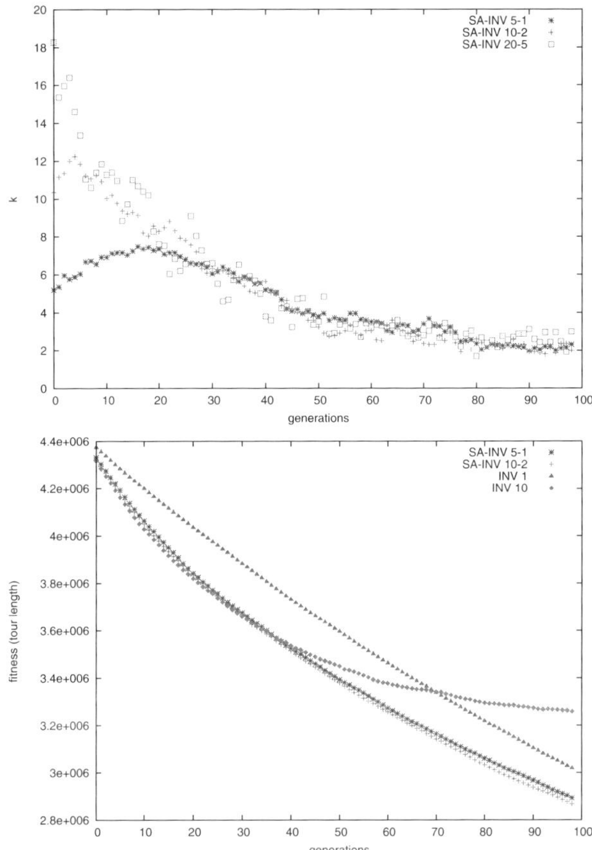

**Fig. 5.5.** SA-INV on problem *gr666*. Upper part: Average development of $k$ over 25 runs. The mutation strength $k$ is spread more broadly than in the previous experiments. Lower part: Average fitness development of SA-INV and INV over 25 runs. SA-INV$_{5-1}$ and SA-INV$_{10-2}$ are the fastest algorithms, at least until $t = 100$. INV$_{10}$ suffers from premature fitness stagnation.

of the instance. There are 666! possible tours and the number of inversions in each generation until successful solutions are produced might be spread broadly around a certain value. Here it is noticeable that the comma selection allows deteriorations. Furthermore, we can observe an increase of $k$ at the beginning of the search, slower than in the previous experiments, but appearing for all three SA-INV variants. Again, we observe the premature fitness stagnation of INV$_{10}$ and the superiority of SA-INV$_{5-1}$ and SA-INV$_{10-2}$, at least until generation $t = 100$.

### 5.4.4  Small TSP Instances

On smaller instances like *ulysses16* we observed principally a similar development of $k$. Within the first generations $k$ is bigger than one. After 10 to 20 generations $k$ almost constantly decreases to 1. Due to comparatively small improvements at the beginning of the search, we skip the presentation of these results. They are comparable to the results on the other TSP instances.

### 5.4.5  Statistical Test

We apply the non-parametric Wilcoxon rank-sum test on the results of the previous sections, in order to prove statistically that the SA-INV variants are significantly better than the best INV variants.

**Wilcoxon Rank-Sum Test**

| | berlin52 | bier127 | gr666 |
|---|---|---|---|
| | SA-INV$_{5-1}$ vs. INV$_{10}$ | SA-INV$_{5-1}$ vs. INV$_1$ | SA-INV$_{10-2}$ vs. INV$_1$ |
| $p_w$ | 0.0001129 | 1.312E-07 | 5.561E-08 |

To this end we compare the results of SA-INV$_{5-1}$ to the results of INV$_{10}$ on instance *berlin52*.[4] The small $p_w$-value of $0.0001129 << 0.05$ leads to a rejection of the hypothesis that both heuristics produce an identical distribution. The median and mean of SA-INV$_{5-1}$ is smaller, hence its results are significantly better. The comparisons between SA-INV$_{5-1}$ and INV$_1$ on *bier127* and SA-INV$_{10-2}$ and INV$_1$ on *gr666* with the small $p_w$-values lead to the same conclusion: the self-adaptive variants produce significantly better results than the standard inversion operator.

## 5.5  The Strategy Bound Problem and SA-INV-c

The self-adaptation of $k$ speeds up the starting phase of the search. At the beginning, more than one inversion mutation in each step leads to considerable improvements. Parameter $k$ decreases constantly, as less and less edge swaps are necessary to improve the solution and not to destroy it. After a certain number of generations $g$, the strategy parameter $k$ is almost constantly one. From now on more than one change of edges leads to a fitness deterioration. But the mutation of strategy variables in the context of the self-adaptive process still produces solutions with $k > 1$ and consequently offspring with potentially worse fitness. The progress of SA-INV slows down and soon INV$_1$ *overtakes* SA-INV concerning the approximation. We call this problem *strategy bound problem*: Whenever the strategy variables reach a bound $b$, the self-adaptive process experiments with

---

[4] The mean of INV$_{10}$ is better than the mean of INV$_1$, the median is worse. A test of INV$_1$ and SA-INV$_{5-1}$ produces $p_w \leq 0.01166$, which is still significant.

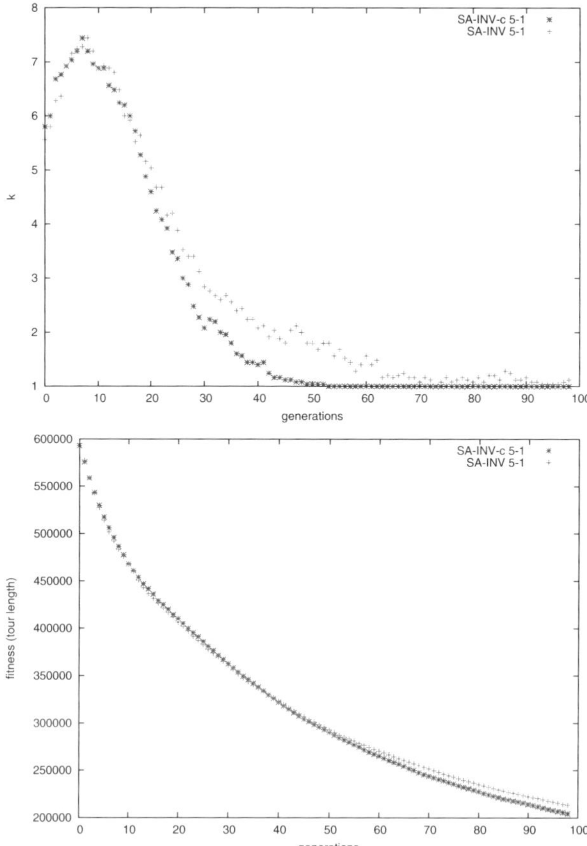

**Fig. 5.6.** Comparison of the mutation strength $k$ (upper part) and the fitness (lower part) of SA-INV-c$_{5-1}$ and SA-INV$_{5-1}$ on problem *bier127* averaged over 25 runs. Once $k$ reaches the region around one it is constantly set to $k = 1$. The fitness development shows the superiority of SA-INV-c in comparison to SA-INV, in particular at later generations.

settings around $b$. But the bound itself may be an optimal setting. All other settings produce potentially worse solutions. Consequently, the success probability is smaller than $p_s$ with the setting $b$. A bound of strategy variables is a phenomenon that occurs almost naturally for *discrete* strategy variables.

In order to overcome this problem we propose the following modification SA-INV-c, which is based on SA-INV. Let $u$ be the number of individuals with $k = 1$. If $u \geq \frac{\lambda}{2}$ we set $k = 1$ constantly and cancel the self-adaptation of $k$ during the following generations. This modification makes sense under the assumption that after $\kappa$ generations the optimal strategy setting is $k = 1$ and for $k > 1$ only worse solutions are produced. We assume that $g$ is reached if the condition $u \geq \frac{\lambda}{2}$ is fulfilled. This bound is reasonable: if at least 50% of the strategy parameters

**Table 5.4.** Experimental comparison of SA-INV-c, SA-INV and INV on problem *bier127* after 100 generations. SA-INV-c achieves the best results for *best, median* and *mean.*

|        | SA-INV-c$_{5-1}$ | SA-INV$_{5-1}$ | INV$_1$ | INV$_{10}$ |
|--------|-----------|-----------|-----------|-----------|
| best   | 190912.96 | 191115.51 | 194403.16 | 402244.68 |
| median | 200495.52 | 214640.67 | 205783.69 | 419211.84 |
| worst  | 235531.40 | 234179.99 | 221001.62 | 447189.49 |
| mean   | **202989.63** | 212523.76 | 206272.24 | 421587.75 |
| dev    | 10327.06  | 9954.37   | 6339.65   | 12164.78  |

*vote* for the setting $k = 1$, we approve this choice for all parameters. We tested the SA-INV-c$_{5-1}$ algorithm experimentally on instance *bier127* and present the results in figure 5.6. The upper part shows the development of mutation strength $k$. As expected once reached the region of one it is constantly set to $k = 1$ to avoid solution destroying mutations. The lower part of the figure illustrates the superiority of SA-INV-c$_{5-1}$ in comparison to SA-INV$_{5-1}$. SA-INV-c$_{5-1}$ is faster, in particular in later phases of the search.

Table 5.4 confirms the results. It shows the fitness after 100 generations on problem *bier127*. SA-INV-c$_{5-1}$ achieves the best results for *best, median* and *mean*. The Wilcoxon rank-sum test reveals that SA-INV-c$_{5-1}$ is significantly better than SA-INV$_{5-1}$ ($p_w \leq 0.0001043$), also better than INV$_1$ ($p_w \leq 0.04781$), and definitely better than INV$_{10}$ ($p_w \leq 1.4E - 9$). SA-INV$_{5-1}$ and INV$_1$ do not produce significantly different solutions. Hence, the experiments show that self-adaptive inversion mutation with the heuristic extension (SA-INV-c) is able to beat standard inversion mutation.

## 5.6  Summary

1. Despite the fact that only very few examples exist, where the self-adaptation of discrete variables for EAs work properly, in this chapter we presented the SA-INV, a successful example for discrete mutation strength self-adaptation.
2. Inversion mutation is a common mutation operator for GAs on the TSP. SA-INV is its self-adaptive variant. The strategy parameter $k$ determines the number of successive applications of inversion mutation, i.e. swaps of edges.
3. Our convergence analysis revealed that an EA with INV or SA-INV converges with probability one to the global optimum after a finite number of iterations.[5]
4. Experiments showed that self-adaptation is able to speed up the optimization, in particular at the beginning of the search. The results have been validated with the Wilcoxon rank-sum test.

---

[5] This statement holds as long as the other genetic operators fulfill the necessary conditions.

5. Parameter $k$ is usually decreasing during the optimization as less and less edge swaps are advantageous for the improvement and not the destruction of solutions. When reaching $k \approx 1$, SA-INV slows down as solutions with $k > 1$ are possible, but deteriorate the optimization. We call this problem *strategy bound problem*. Like in the case of step sizes for ES too big mutations destroy solutions and hinder the evolutionary process from approximating the optimum.

6. To overcome the *strategy bound problem* in later search phases, we introduced the SA-INV-c, a heuristic that sets $k = 1$ constantly if the condition $u \geq \frac{\lambda}{2}$ is fulfilled. The modification improves the performance of SA-INV as the small mutation strength $k = 1$ is essential for successful solutions in the course of the optimization process.

# 6 Self-Adaptive Crossover

*Just as a child creates magnificent fortresses through the arrangement of simple blocks, so does a genetic algorithm seek near optimal performance through the juxtaposition of short, low-order, high-performance schemata, or building blocks.*
Goldberg about the building block hypothesis [47], p. 41.

Crossover plays a controversially discussed role within EAs. The building block hypothesis (BBH) assumes that crossover combines *different* useful blocks of parents. The genetic repair effect (GR) hypothesis assumes that *common* properties of parental solutions are mixed. Most of today's crossover operators do not exhibit adaptive properties, but are either completely random or fixed. In this chapter self-adaptive crossover operators for string and combinatorial representations are introduced. For GAs in string representation self-adaptive crossover means the automatic adaptation of crossover points. With self-adaptive crossover we take a step towards exploiting the structure of the problem automatically, i.e. finding the building blocks or common features self-adaptively. The features we prose to control self-adaptively are crossover points for string representations and recombination factors for ES. A success of self-adaptive crossover would be a hint for the existence of building blocks. A failure would not prove that building blocks do not exist, but could either be a hint for their absence or for the fact that they cannot be identified with self-adaptive crossover points.

In section 6.1 we discuss the role of crossover and introduce the self-adaptive crossover concept. Section 6.2 introduces self-adaptive n-point crossover with an experimental analysis. We test the SA crossover operators for string representations on the problems onemax, ackley and 3-SAT. Partially mapped crossover (PMX) is a crossover operator for combinatorial search domains. In section 6.3 we introduce a self-adaptive variant of PMX. Afterwards, section 6.4 analyzes self-adaptive recombination for ES. Although the experimental results of this chapter are not very encouraging, the crossover point optimization EA of section 6.5 shows that *optimal* crossover points exist.

## 6.1 The Self-Adaptive Crossover Concept

There are two reasons why self-adaptation for crossover is not a simple undertaking. First, the set of problems where crossover might be beneficial is small

O. Kramer: Self-Adaptive Heuristics for Evolutionary Computation, SCI 147, pp. 97–113, 2008.
springerlink.com

[39]. Second and this argument is closely connected to the previous one, the link between crossover strategy parameters and their impact on the fitness is smaller than necessary for successful self-adaptation [8].

### 6.1.1   The Role of Crossover - Building Blocks or Genetic Repair?

The assumption of the BBH by Goldberg [47] and Holland [59] is that different good building blocks from a parental population are mixed together during crossover in order to combine good properties of the parents and inherit these to the offspring. Mitchell et al. [95] report that finding test functions to support this hypothesis is surprisingly difficult. Jansen and Wegener [62] constructed an artificial test function where one-point crossover improves the time complexity. But it is still unexplained, whether crossover helps generally on non-artificial test functions.

In contrast to the BBH, the GR hypothesis claims that recombination does not let the *different*, but the *common* features flow into the offspring [14]. According to this theory recombination extracts the similarities from the parents. Beyer [91] assumes that similar corresponding components in the parental genomes carry a higher probability of being beneficial. He states that the best a crossover operator can do is to conserve these components. Beyer [16] states that recombination itself usually has no benefit without mutation, but is only useful with a high population variance. This high variance can be a result of a high mutation strength or random initialization. The latter is the source for necessary variance of GA implementations that do not use any mutations. In other words: according to the GR hypothesis mutations are absolutely necessary for the exploration aspect of the search. Selection guides the search by choosing the useful mutations. Up this stage BBH and GR agree. But recombination only reduces the statistically uncorrelated parts of the mutations. This assumes that these similarities are statistically the beneficial ones. From his hypothesis Beyer [16] proposed the following principles for the design of crossover operators:

- *Problem-specificness.* Recombination should be constructed problem-specific to make the extraction of the statistical similarities possible.
- *Absence of biases.* the probability for each individual to contribute to the offspring should be equal.
- The quality of the statistical estimation increases with the number of parents $\rho$ participating. From this point of view, recombination conflicts with selection and postulates $\rho = \lambda = \mu$, i.e. no selection pressure. Of course, in practice $\mu$ has to be chosen problem specific.
- *Permanence.* Recombination should always be used, i.e. $p_c = 1.0$. According to Beyer [16] the empirical results for smaller values are due to improperly chosen mutation strengths.

Beyer states that these criteria do not identify recombination operators completely. His results are theoretically valid on the sphere function. On other, more complex functions, the principles may not be feasible possibly.

## 6.1.2   Preliminary Work

Self-adaptation for crossover played only a marginal role for EA research in the last decades. Instead, most research concentrated on mutation operators. One reason for this might be the historical development as it was first introduced for ES and EP, where crossover was almost neglected at the beginning [91]. The first attempt to integrate self-adaptation into crossover was *punctuated crossover* by Schaffer and Morishima [128]. Punctuated crossover adapts the crossover points for bit string representations and was experimentally proved better than standard one-point crossover. A strategy parameter vector with the same length as the objective variable vector is added to the individual's representation. It is filled with ones, the so called punctuation marks, wherever a crossover point exists. Punctuation marks from both parents indicate a change from which parent the next part of the genome has to be copied to the offspring. This self-adaptive variant outperforms a canonical GA on four of five De Jong's test functions. Two interesting observations were made: First, some bit positions tended to accumulate more punctuation marks than others and these accumulations changed over time. Second, the number of events, when nonidentical gene segments are swapped between parents while offspring different from its parents is produced is constant and depends on the representation length. Spears [147] does not hold self-adaptation responsible for the success of punctuated crossover, but the usage of more than one crossover points. He proposed the one-bit self-adaptive crossover, which makes use of a bit indicating the crossover type. This technique is similar to our self-adaptive mutation operator selection of section 4.6. The bit switches between uniform and two-point crossover. From his tests on N-peak problems Spears derived the conclusion that the key feature of his self-adaptation experiment is not the possibility to select the best operator, but the degree of freedom for the EA to choose among two operators during the search. This kind of self-adaptation only concerns the selection of a crossover type, but not its structure like in our approach. Maruo et al. introduced an approach, where self-adaptive crossover means the control of the crossover probability $p_c$ [88].

Another technique refers to the identification of building blocks, the LEGO (linkage evolving genetic operator) algorithm [143]. Each gene exhibits two bits indicating the linkage to neighboring genes which is only possible if the same bits are set. New candidate solutions are created from left to right conducting tournaments between parental blocks. Meyer-Nieberg and Beyer [91] point out that even the standard 1- or n-point crossover operators for bit string representations exhibit properties of a self-adaptive mutation operator. Bit positions which are common in both parents are transferred to the offspring individuals while other positions are filled randomly. This assumption is consistent with the GR hypothesis and in contrast to the BBH.

## 6.1.3   Adaptation of the Crossover Structure

The concept of self-adaptation is a step into the direction of parameter independent EC. Self-adaptation is usually associated with the step size control of ES.

It was originally introduced by Rechenberg [114] and Schwefel [132], later by Fogel [42] for EP. Self-adaptive algorithms are characterized by the integration of strategy parameters, which influence the genetic operators, into the individual's chromosomes. These parameters undergo genetic variation and are involved into the evolutionary optimization process. For detailed descriptions and definitions we refer to chapter 3.

Crossover is the main genetic operator besides mutation and selection. Its role has not been satisfactorily explained. Is it just creating noise similar to mutation or are there really building blocks, structurally connected parts of the solution useful to exchange during evolution? And where are the borders of these building blocks, the best positions for crossover points? Do they really exist, and, if so, are they dynamically changing during the optimization process or are they constant? We think that these questions cannot be answered in general, but are problem dependent. In the following we attempt to use the concept of self-adaptation to let the evolution itself identify the structural parameters of crossover. It was our hope that this exploitation aspect of crossover is a source of effectivity and efficiency improvements. For string representations these parameters mainly concern the crossover points.

In this chapter the designs for self-adaptive crossover operators are proposed. For binary and integer values the string representation is commonly used. For these kinds of representations intuitive crossover operators exist. Let $\mathbf{p}^1, \ldots, \mathbf{p}^\rho$ be $\rho$ parents, each consisting of objective variables $O = (p_1, \ldots, p_N)$. We extend the genome by strategy variables $\Sigma = (\Lambda_1, \ldots, \Lambda_n)$, which encode self-adaptive crossover features. Each individual consists of $\mathbf{p}_i = (O, \Sigma)$, the same holds for offspring individuals $\mathbf{o}_i$.

## Crossover Point Selection Schemes

Consider $\rho$ parents are used during recombination. It has to be specified, which crossover point set $\Sigma^\zeta$ of the $\rho$ parents are used. We propose the following heuristic to determine $\Sigma^\zeta$ :

- *Best inheritance.* In this mode the strategy set $\Sigma^\zeta$ of the best of the randomly chosen $\rho$ parents is used and inherited to all offspring solutions.

$$\zeta = argmax_{j,\ 1 \leq j \leq \rho} f(\mathbf{p}^j) \tag{6.1}$$

For the sake of completeness we introduce the variations *dominant* and *intermediate inheritance*:

- *Dominant inheritance.* A parent $\mathbf{p}^\zeta$ is randomly chosen from the $\rho$ parents with uniform distribution. Its strategy set $\Sigma^\zeta$ is taken.
- *Intermediate inheritance.* For each component of the $\rho$ strategy sets an intermediate strategy set $\Sigma$ is computed by

$$\Lambda_i = \frac{\sum_{j=1}^{\rho} \Lambda_k^j}{\rho}. \tag{6.2}$$

As the components of $\Sigma$ are usually integers, i.e. discrete crossover points, rounding procedures discretize $\Lambda_i$ afterwards.

For our experiments we used the *best inheritance* scheme. In our analysis no better results could be achieved with the other two variants.

**Strategy Variable Mutation**

Self-adaptation is only possible if the strategy variables are modified by the genetic operators. The mutation of the crossover points depends on the underlying representation. If encoded binary we propose to use classic bit-flip mutation resulting in a geometric distribution. For continuous search domains we propose to use Gaussian mutation, i.e. the addition of a random Gaussian distributed value. We use meta-EP mutation with step size $\sigma$

$$\Lambda' = \Lambda + \sigma \cdot \mathcal{N}(0,1) \text{ with } \Lambda' \in [1, l-1]. \tag{6.3}$$

The step size $\sigma$ has to be rounded and adjusted according to the length of the representation. As an alternative log-normal mutation may be used:

$$\Lambda' = \Lambda \cdot e^{\sigma \cdot \mathcal{N}(0,1)} \text{ with } \Lambda' \in [1, l-1] \tag{6.4}$$

Log-normal mutation is used for the step size mutation of ES, see chapter 4.

## 6.2 Self-Adaptive n-Point Crossover

The idea of self-adaptive n-point crossover is to divide the genome of the parents into n+1 parts and combine them with the aim of identifying useful building blocks. If they exist, it will make sense to set the crossover points to the borders of the blocks in order to combine them, at least not to destroy the blocks. In the following we introduce the self-adaptive operators SA-1-point, SA-n-point and self-adaptive multi parent genetic algorithms (MPGAs).

### 6.2.1 SA-1-Point

The variant 1-point crossover works as follows: A crossover point $\Lambda$ is chosen randomly with uniform distribution. The left part of the bit-string from the first parent is combined with the right part of the second parent and vice versa for a second solution. For the self-adaptive extension of 1-point crossover (SA-1-point) the strategy part of the chromosome is extended by an integer value $\Sigma = (\Lambda)$ with $\Lambda \in [1, l-1]$ determining the crossover point. Given two parents

$$\mathbf{p}^1 = (p_1^1, \ldots, p_l^1, \Lambda^1) \tag{6.5}$$

and

$$\mathbf{p}^2 = (p_1^2, \ldots, p_l^2, \Lambda^2) \tag{6.6}$$

self-adaptive 1-point crossover selects the crossover point $\Lambda^\varsigma$ of one of the two parents according to the proposed selection schemes and creates the two children

$$\mathbf{o}^1 = (p_1^1, \ldots, p_{\Lambda^\varsigma}^1, p_{\Lambda^\varsigma+1}^2, \ldots, p_l^2, \Lambda^\varsigma) \tag{6.7}$$

and

$$\mathbf{o}^2 = (p_1^2, \ldots, p_{\Lambda^\varsigma}^2, p_{\Lambda^\varsigma+1}^1, \ldots, p_l^1, \Lambda^\varsigma). \tag{6.8}$$

The crossover point $\Lambda_k$ at locus $k$ is the position between element $k$ and $k+1$. We proposed three schemes for the decision, which crossover point to take in section 6.1.3. The strategy part $\Lambda^\varsigma$ may be encoded as an integer or in binary representation as well. In order to enable self-adaptation the crossover point $\Lambda^\varsigma$ has to be mutated. For this purpose meta-EP or log-normal mutation and the usage of rounding procedures is proposed, see section 6.1.3.

---

**Experimental settings**

| | |
|---|---|
| Population model | generational, $\mu = 150$ |
| Mutation type | bit flipping, $p_m = 0.001$ (onemax, sphere), |
| | $p_m = 0.005$ (ackley, 3-SAT) |
| Strategy mutation | meta-EP |
| Crossover type | 1-point, SA-1-point, $p_c = 1.0$ |
| Selection type | generational model |
| Initialization | random bits |
| Termination | optimum found |
| Runs | 50 |

---

The bigger mutation strength of $p_m = 0.005$ on the problems ackley and 3-SAT showed better results than $p_m = 0.001$, which was better on onemax and sphere. We used gray coding for the numerical functions. Table 6.1 shows the

**Table 6.1.** Experimental comparison of 1-point and SA-1-point crossover on onemax, sphere, ackley and 3-SAT. The figures show the number of generations until the optimum is found. No significant superiority of any of the two 1-point variants can be detected.

| | best | worst | mean | dev |
|---|---|---|---|---|
| *onemax* | | | | |
| 1-point | 16 | 93 | 28.88 | 2.23 |
| SA-1-point | 13 | 71 | **27.68** | 1.97 |
| *sphere* | | | | |
| 1-point | 11 | 102 | 42.25 | 4.99 |
| SA-1-point | 16 | 89 | **40.33** | 4.10 |
| *ackley* | | | | |
| 1-point | 13 | 438 | 63.50 | 86.83 |
| SA-1-point | 15 | 455 | **57.01** | 58.95 |
| *3-SAT* | | | | |
| 1-point | 15 | 491 | 63.68 | 291 |
| SA-1-point | 7 | 498 | **62.72** | 264 |

experimental results of SA-1-point on the problems onemax, sphere[1], ackley and 3-SAT. The table shows the number of generations until the optimum is found. Although in all experiments the mean of SA-1-point lies under the mean of 1-point crossover, the standard deviation is too high to assert statistical significance. SA-1-point was able to find the optimum of 3-SAT in only 7 generations, which is very fast. But it also needs the longest time with its worst result. We observe no significant improvement of the self-adaptive variants in comparison to the standard operators.

### 6.2.2   SA-n-Point

For n-point crossover the representation is broken into more than two segments and assembled alternately from both parents. For the self-adaptive extension (SA-n-point) the n crossover points are saved in the genome as strategy variables $\Sigma = (\Lambda_1, \ldots, \Lambda_n)$, sorted by size $\Lambda_1 < \Lambda_2 < \ldots < \Lambda_n$. Given two parents $\mathbf{p}^1 = (p_1^1, \ldots, p_l^1, \Lambda_1^1, \ldots, \Lambda_n^1)$ and $\mathbf{p}^2 = (p_1^2, \ldots, p_l^2, \Lambda_1^2, \ldots, \Lambda_n^2)$, SA-n-point crossover selects a crossover point set $\Sigma^\zeta$ to create the new offspring $\mathbf{o}^1$ and $\mathbf{o}^2$ with

$$o_i^1 = \begin{cases} p_i^1, & \Lambda_m^\zeta < i \le \Lambda_{m+1}^\zeta \quad m \text{ even} \\ p_i^2, & \Lambda_m^\zeta < i \le \Lambda_{m+1}^\zeta \quad m \text{ odd} \end{cases} \tag{6.9}$$

and

$$o_i^2 = \begin{cases} p_i^2, & \Lambda_m^\zeta < i \le \Lambda_{m+1}^\zeta \quad m \text{ odd} \\ p_i^1, & \Lambda_m^\zeta < i \le \Lambda_{m+1}^\zeta \quad m \text{ even} \end{cases} \tag{6.10}$$

with $\Lambda_0^\zeta = 0$ and $\Lambda_{n+1}^\zeta = l$ The strategy part $\Sigma^\zeta$ has to be mutated, sorted and included into each offspring genome afterwards. In order to evaluate the success of SA-n-point crossover, we take the test problems of the previous section and test SA-5-point with five crossover points. Table 6.2 shows the results, using the settings of the previous experiments. Similar to the previous series of experiments we cannot observe a significant improvement of self-adaptive crossover. Neither on the unimodal functions (onemax and sphere) nor on the highly multimodal SAT-problem the self-adaptation of crossover points seems to be effective. Hence, these experiments call the possibility that building blocks can be identified automatically with self-adaptation into question. But if we compare the results to 1-point crossover, we observe that the increase of crossover points achieves improvements. This observation mainly concerns the mean and the standard deviation.

### 6.2.3   Self-Adaptive Uniform and Multi Parent Crossover

For the sake of completeness we now equip uniform crossover and multi parent genetic algorithms (MPGAs) with self-adaptation, but without presenting an

---

[1] in binary representation

**Table 6.2.** Experimental comparison of 5-point and SA-5-point crossover on onemax, sphere, ackley and 3-SAT. The figures show the number of generations to reach the optimum. Again, no consistent picture can be drawn concerning the superiority of any variant.

|  | best | worst | mean | dev |
|---|---|---|---|---|
| *onemax* | | | | |
| 5-point | 12 | 32 | **17.25** | 0.12 |
| SA-5-point | 12 | 44 | 18.07 | 0.22 |
| *sphere* | | | | |
| 5-point | 11 | 83 | **26.03** | 2.39 |
| SA-5-point | 10 | 83 | 27.23 | 2.73 |
| *ackley* | | | | |
| 5-point | 14 | 412 | **50.44** | 58.96 |
| SA-5-point | 16 | 494 | 52.7 | 95.11 |
| *3-SAT* | | | | |
| 5-point | 11 | 467 | 46.57 | 166 |
| SA-5-point | 14 | 499 | **42.96** | 152 |

experimental analysis. Again, given two parents $\mathbf{p}^1$ and $\mathbf{p}^2$, the genes are taken from one of the two parents randomly, in most cases with probability $p = 0.5$. This becomes

$$o_i = p_i^j, \quad j \in \mathrm{random}\{1,2\}. \tag{6.11}$$

To equip uniform crossover with self-adaptation (SA uniform), we introduce a strategy set $\Sigma = (\Lambda_1, \ldots, \Lambda_l)$ with $\Lambda_i \in \{0, 1\}$ determining from which parents to take the genes

$$o_i = p_i^{\Lambda_i^\zeta} \tag{6.12}$$

It is equivalent to dSAR, presented in section 6.4.3. We point out the opinion that self-adaptive uniform crossover could make sense, in particular when the involved parents come from different mating pools, from different subpopulations or exhibit maximal Euclidean distance.

In nature, genetic material is exclusively exchanged between two parents. But the artificial modeling allows to acquit from this restriction. Thus, MPGAs allow more than two parents to participate in crossover simultaneously. An analysis of MPGAs can be found in the paper of Ting [153]. The idea of diagonal crossover is, similarly to N-point crossover, the division of the parental genomes into N+1 parts and the alternate assembly of the offspring. Self-adaptive diagonal crossover (SA-diagonal) works as follows: given $\rho$, typically $\rho = n + 1$, parents $\mathbf{p}^1, \ldots, \mathbf{p}^\rho$ and $n$ crossover points $\Lambda_1^\zeta, \ldots, \Lambda_{n-1}^\zeta \in \{1, \ldots, l - 1\}$ with

$\Lambda_1^\zeta < \Lambda_2^\zeta < \ldots < \Lambda_{n-1}^\zeta$, which are saved in the strategy part $\Sigma^\zeta$, the offspring is produced in the following way:

$$o_i = p_i^{j-1 \text{ modulo } n} \quad \text{for } \Lambda_j^\zeta < i \le \Lambda_{j+1}^\zeta \tag{6.13}$$

with $\Lambda_0^\zeta = 1$ and $\Lambda_{n+1}^\zeta = l$. U-Scan is the multi-parent variant of uniform crossover. The idea is to assign each of the offspring's gene position randomly with one of the parent's genes at the same position. Self-adaptive U-Scan (SA-U-Scan) saves the parent $j$, from which to take the gene for every position $i$. Hence, $\Sigma^\zeta$ is a N-dimensional vector of integers $\Sigma^\zeta = (\Lambda_1^\zeta, \ldots \Lambda_N^\zeta)$ with $1 \le \Lambda^\zeta \le \rho$. SA-U-Scan produces new solutions as follows:

$$o_i = p_i^{\Lambda_i^\zeta} \tag{6.14}$$

OB-scan performs a majority decision, the most frequent gene of all parents at a position determines the gene of the offspring at the same position. Thus, this operator cannot be equipped with self-adaptation.

## 6.3   Self-Adaptive Partially Mapped Crossover

Partially mapped crossover (PMX) was designed for TSP by Goldberg and Lingle [48] and enhanced by Whitley [159]. The PMX works as follows:

- Choose two parents $\mathbf{p}^1$ and $\mathbf{p}^2$,
- choose two crossover points $\Lambda_1$ and $\Lambda_2$,
- copy the segment between $\Lambda_1$ and $\Lambda_2$ from $\mathbf{p}^1$ to offspring $\mathbf{o}^1$,
- put all genes of $\mathbf{p}^2$ in this segment which have not been copied into set $\mathcal{L}$,
- for each element $l$ of $\mathcal{L}$ look for the corresponding locus in parent $\mathbf{p}^1$ and copy $l$ to this position if free. If this position is not free, recursively repeat the process until a free position can be found,
- fill the empty loci with genes from corresponding loci of $\mathbf{p}^2$.

We extend the PMX to self-adaptive partially mapped crossover (SA-PMX). SA-PMX keeps the crossover points $\Lambda_1$ and $\Lambda_2$ in the strategy part $\Sigma$ and evolves them during the optimization process [74]. We use the *best inheritance* heuristic to determine the strategy set $\Sigma^\zeta$, i.e. $\zeta = argmax_{j,\ 1 \le j \le \rho} f(\mathbf{p}^j)$. Self-adaptation is only possible if the strategy variables are modified with the genetic operators. We propose to use meta-EP mutation for the crossover points.

| Experimental settings | |
|---|---|
| Population model | steady state, (100,200) |
| Mutation type | inversion mutation, $\sigma = 3, 5, 25$ |
| Crossover type | SA-PMX |
| Selection type | fitness proportional |
| Initialization | random |
| Termination | 500,2000,3000 generations |
| Runs | 25 |

**Table 6.3.** Analysis of standard PMX and SA-PMX on three TSP instances. The figures show the tour length after the maximum number of generations. Although SA-PMX achieves slightly better results, the standard deviation is too high to speak of significant superiority. But the results of SA-PMX are better than the results of $\text{PMX}_c$ with constant crossover points.

|  | best | mean | dev | worst | optimum |
|---|---|---|---|---|---|
| *ulsysses16* | | | | | |
| PMX | 6859.0 | 6866.8 | 10.6 | 6950.0 | 6859.0 |
| $\text{PMX}_c$ | 6859.0 | 6869.6 | 15.0 | 6950.0 | |
| SA-PMX | **6859.0** | **6864.3** | **6.3** | **6875.0** | |
| *berlin52* | | | | | |
| PMX | **7642.1** | 8366.4 | 270.9 | 9297.6 | 7542.0 |
| $\text{PMX}_c$ | 7887.3 | 8539.6 | 269.0 | 9335.7 | |
| SA-PMX | 7777.3 | **8304.2** | **249.5** | **8921.0** | |
| *bier127* | | | | | |
| PMX | **123765.0** | 133600.4 | **3399.7** | 141157.7 | 118282 |
| $\text{PMX}_c$ | 131124.4 | 137679.4 | 3592.6 | 143680.4 | |
| SA-PMX | 125620.9 | **133594.2** | 3426.5 | **140236.1** | |

We tested SA-PMX on three TSP instances [74] from the TSPlib of Reinelt [117]. For the experiments we use inversion mutation[2] with $p_m = 0.1$, crossover probability of $p_c = 1.0$, population size 100 and fitness proportional parent selection. Besides PMX we also tested a variant with fixed crossover points at $\Lambda_1 \approx \frac{N}{3}$ and $\Lambda_2 \approx \frac{2N}{3}$. On TSP instance *ulysses16* [$\sigma = 3$, termination after 500 generations] all variants reached the optimum in the best run, but SA-PMX achieved the best average and worst result with the lowest standard deviation. On problem *berlin52* [$\sigma = 5$, termination after 2000 generations] PMX achieved the shortest but not optimal tour length of all runs. SA-PMX succeeded considering the average and the worst tour. On *bier127* [$\sigma = 25$, termination after 3000 generations] the PMX attained the best overall tour length, while SA-PMX achieved the best average and worst as well as the lowest standard deviation. But due to the standard deviation we have to point out that we cannot speak of significant results: The behavior of the algorithm with regard to the problem is *not* changed by the introduced self-adaptation. In every case the variant $\text{PMX}_c$ with fixed crossover points achieved the worst results. Obviously, the randomness is essential for the success of crossover.

To summarize the experimental results, a slight but not statistically significant improvement can be reported with self-adaptation of crossover points. No consistent picture concerning the superiority of SA-PMX can be drawn. Figure 6.1 shows a comparison of both crossover points of PMX (left part) and both of SA-PMX (right part) averaged in each generation on problem *bier127* during 3000 generations. The crossover points of PMX are stable, oscillating around the

---

[2] Here, we do not use the self-adaptive variant SA-INV introduced in chapter 5.

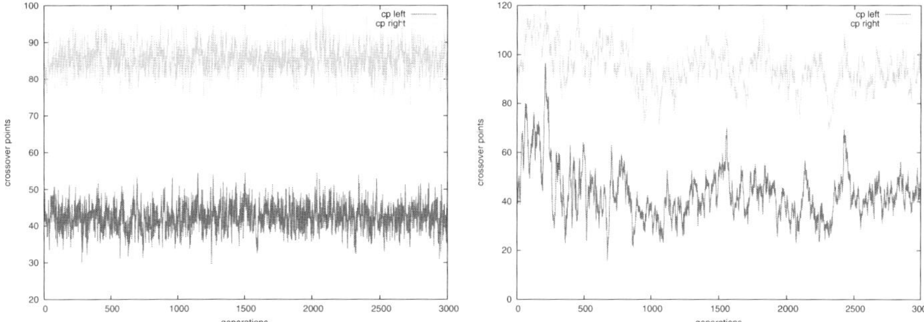

**Fig. 6.1.** Comparison of both randomized crossover points of PMX (left) and both self-adaptive crossover points of SA-PMX (right) on problem *bier127* during 3000 generations. Each figure shows the two **average** crossover points of each generation. Left: The crossover points of PMX are stable, oscillating around the same values during the whole optimization. Right: The crossover points of SA-PMX are less oscillating, but drifting into changing directions for many generations.

same values during the whole optimization. The crossover points of SA-PMX are less oscillating, but drifting into changing directions for many generations. The question stays open, whether we can derive from these observations that a meaningful adaptation of crossover points occurs and building blocks are identified. In particular, we see the challenge to tighten the link between strategy parameter adaptation and fitness. We we hold an undersized link responsible for the shortcomings in comparison to self-adaptation of step sizes. But as already pointed out in chapter 3, a the convergence of strategy parameters is not a necessary condition for self-adaptation.

## 6.4 Self-Adaptive Recombination for Evolution Strategies (SAR)

We let the concept of self-adaptation discover the beneficial settings for the various types of crossover parameters of the different operator types. Concerning its benefits we can argue from different directions. From the point of view of the BBH this approach is supposed to identify the useful building blocks. From the point of view of the GR hypothesis self-adaptive crossover can automatically increase the probability of common successful parental genes. In the following we will recapitulate the two main crossover operators for ES, intermediate and dominant crossover. Afterwards we introduce a more general operator, self-adaptive recombination (SAR).

### 6.4.1   Intermediate and Dominant Recombination

Two main variants of recombination for ES exist. *Intermediate* crossover produces new offspring solutions $\mathbf{o} = (o_1, \ldots, o_N)$ by calculating the arithmetic median of $\rho$ randomly selected parents $\mathbf{p}^k$ with $1 \leq k \leq \rho$ and $\mathbf{p}^k = (p_1^k, \ldots, p_N^k)$:

$$o_i := \frac{1}{\rho} \sum_{k=1}^{\rho} p_i^k \tag{6.15}$$

The equation yields the centroid of the selected parents. For discrete representations rounding procedures have to be used. *Dominant* crossover chooses each component from one of the $\rho$ parents randomly with uniform distribution:

$$o_i := p_i^k \text{ with } k := \text{Random } \{1, \ldots, \rho\} \tag{6.16}$$

Dominant crossover became famous as *uniform* crossover in the field of GAs. In the case of $\rho = \mu$ it is also called global discrete recombination.

### 6.4.2   Self-Adaptive Recombination

The idea of our SAR is to *morph* between intermediate and dominant recombination for two parents ($\rho = 2$) self-adaptively. We introduce a recombination coefficient vector $\Xi$, which determines the amount of information inherited by each parent:

$$\Xi = \nu_i \text{ with } \nu_i \in [0, 1] \quad \text{ and } 1 \leq i \leq N \tag{6.17}$$

Hence, each individual $\mathbf{a}$ now consists of a vector

$$\mathbf{a} = (x_1, \ldots, x_N, \sigma_1, \ldots, \sigma_N, \nu_1, \ldots \nu_N) \tag{6.18}$$

During recombination each component $o_i$ of the objective part of the offspring $\mathbf{o} = (o_1, \ldots, o_N)$ is produced in the following way given the two randomly selected parents $\mathbf{p}^1$ and $\mathbf{p}^2$:

$$o_i = \nu_i^{\zeta} \cdot p_i^1 + (1 - \nu_i^{\zeta}) \cdot p_i^2 \tag{6.19}$$

with the strategy variables from parent $\zeta$. In order to enable self-adaptation, the recombination coefficients have to be mutated. We propose to mutate the recombination coefficients with Gaussian meta-EP mutation

$$\nu' = \nu + \gamma \cdot \mathcal{N}(0, 1) \tag{6.20}$$

with $\gamma \approx 0.1$, similar to the mutation of angles [132] and bias coefficients, see section 4. We use the *best* scheme proposed in section 6.1.3 to determine the parent $\zeta$.

### 6.4.3  SAR Variants

We propose two obvious variants of the SAR operator. The variant individual SAR (**iSAR**) makes use of only one self-adaptive parameter $\nu \in [0, 1]$ on *individual* and not component level. Hence, every individual **a** is now a vector

$$\mathbf{a} = (x_1, \ldots, x_N, \sigma_1, \ldots, \sigma_N, \nu) \tag{6.21}$$

with only one recombination strategy variable. Similar to SAR, a new offspring solution **o** is now produced with

$$o_i = \nu \cdot p_i^1 + (1 - \nu) \cdot p_i^2 \tag{6.22}$$

This variant reduces the self-adaptation effort, but does not allow a component level adaptation.

The morphing between dominant and intermediate crossover is *continuous* for the SAR operator. In order to switch discretely between both variants we introduce the discrete SAR (**dSAR**), which uses an N-dimensional strategy vector of bits.

$$\Xi = \nu_i \text{ with } \nu_i \in \{0, 1\} \text{ and } 1 \leq i \leq N \tag{6.23}$$

These bits determine, which recombination type to use. We define 0 for intermediate and 1 for discrete recombination. Hence, **o** is produced by

$$o_i := \begin{cases} \frac{1}{2}p_i^1 + \frac{1}{2}p_i^2 & \text{if } \nu_i = 0 \\ p_i^k \text{ with } k := \text{Random } \{1, 2\} & \text{if } \nu_i = 1 \end{cases} \tag{6.24}$$

From the EDA point of view, see section 3.7, SAR controls the distances of particular distributions $\mathcal{M}_i$ with $1 \leq i \leq \mu$ of the mixture distribution $\mathcal{M}$.

### 6.4.4  Experimental Analysis

We tested the introduced variants on various typical test functions:

- sphere ($N = 10$),
- rastrigin ($N = 10$),
- rosenbrock with noise in fitness ($N = 10$),
- sharp ridge ($N = 10$),
- and the constrained problem 2.40 ($N = 5$).

Now we present a comparison of the proposed SAR variants with intermediate (int) and dominant (dom) recombination. We use the following experimental settings.

| Experimental settings | |
| --- | --- |
| Population model | (15,100) |
| Mutation type | standard |
| Crossover type | int, dom, SAR, dSAR, $\rho = 2$ |
| Selection type | comma |
| Initialization | [-100,100] |
| Termination | 1000 generations, rosenbrock with noise: 2000 generations |
| Runs | 25 |

Table 6.4 summarizes the experimental results. On the sphere function intermediate recombination is much better than dominant recombination. SAR and iSAR are as good as intermediate recombination, dSAR even slightly worse than dominant recombination. On rastrigin SAR and also iSAR are slightly, but not significantly better than intermediate recombination, see the next Wilcoxon rank-sum test. Intermediate recombination reaches the optimum in 9 of the 25 runs, SAR and iSAR each find the optimum in 10 runs. Dominant and dSAR

**Table 6.4.** Experimental comparison of intermediate, dominant recombination, SAR and dSAR on five selected continuous test functions. SAR is as good as intermediate recombination, but in particular not statistically worse, see the Wilcoxon rank-sum test. On the sharp ridge dominant recombination and dSAR are superior to SAR and int.

| | int | dom | SAR | iSAR | dSAR |
| --- | --- | --- | --- | --- | --- |
| *sphere* | | | | | |
| best | 2.36E-150 | 3.76E-128 | 1.78E-150 | 3.25E-150 | 2.54E-123 |
| worst | 3.60E-141 | 1.12E-109 | 1.52E-140 | 1.10E-141 | 1.70E-108 |
| mean | 2.07E-142 | 4.47E-111 | 6.12E-142 | 4.87E-143 | 6.96E-110 |
| dev | 7.36E-142 | 2.23E-110 | 3.03E-141 | 2.20E-142 | 3.39E-93 |
| *rastrigin* | | | | | |
| best | 0.00 | 1.99 | 0.00 | 0.00 | 4.97 |
| worst | 2.98 | 19.89 | 1.98 | 3.97 | 22.88 |
| mean | 1.15 | 7.64 | 0.79 | 0.95 | 11.42 |
| dev | 1.13 | 4.81 | 0.75 | 1.05 | 5.65 |
| *rosenbrock noise* | | | | | |
| best | 1.31 | 1.75 | 1.26 | 2.18 | 1.99 |
| worst | 19.05 | 8310.17 | 385.20 | 9.41 | 4072.36 |
| mean | 5.76 | 683.08 | 28.30 | 5.43 | 429.60 |
| dev | 3.06 | 2181.75 | 82.70 | 1.23 | 1020.40 |
| *sharp ridge* | | | | | |
| best | -6.29E+192 | -3.51E+248 | -8.13E+187 | -1.84E+190 | -1.62E+242 |
| worst | -1.46E+175 | -7.84E+217 | -1.27E+174 | -2.29E+171 | -3.67E+202 |
| mean | -2.52E+191 | -1.40E+247 | -5.63E+186 | -7.63E+188 | -1.10E+241 |
| 2.40 | | | | | |
| best | -4954.77 | -4941.65 | -4956.74 | -4932.08 | -4950.60 |
| worst | -4099.45 | -3835.10 | -3994.14 | -4019.12 | -4174.98 |
| mean | -4626.94 | -4585.72 | -4569.26 | -4639.31 | -4671.73 |
| dev | 260.12 | 319.05 | 281.91 | 241.60 | 239.13 |

are again a bit worse. On rosenbrock with noise intermediate recombination is as good as iSAR, SAR is almost comparable. The variant dSAR produces outliers, which deteriorate the results. On the ridge functions the situation changes considerably. Dominant crossover becomes the best recombination operator, while dSAR achieves a similar approximation quality. The latter *selects* dominant crossover self-adaptively. Intermediate recombination, SAR and iSAR climb the ridge comparatively slowly. On the constrained problem 2.40 the choice of the recombination operator does not have a major influence. No variant is able to find the optimum. To summarize, intermediate recombination, SAR and iSAR show a similar behavior on the test functions. The next statistical test shows that no significant difference between SAR and intermediate can be reported. Hence, we cannot recommend any variant. The discrete switch between intermediate and dominant crossover dSAR is comparatively weak. Dominant recombination and dSAR are appropriate for ridge functions.

**Wilcoxon Rank-Sum Test**

|  | sphere | rastrigin | roseNoise | sharp ridge | 2.40 |
| --- | --- | --- | --- | --- | --- |
| $p_w$, int vs. SAR | 0.6837 | 0.3879 | 0.2072 | 0.1683 | 0.5737 |

A Wilcoxon rank-sum test is conducted to examine statistical differences between intermediate recombination and SAR. The high $p_w$-values ($p_w > 0.05$) show that there is no statistical indication for the superiority of neither intermediate recombination nor SAR. We observe relatively small values on rosenbrock with noise ($p_w \leq 0.2072$) and on the sharp ridge ($p_w \leq 0.1683$). But this does not allow a significant statement, neither about a superiority of SAR on rosenbrock with noise nor of intermediate recombination on the sharp ridge.

## 6.5 Crossover Point Optimization

As the self-adaptation of crossover points does not lead to any significant improvement, we propose to find optimal crossover points with an optimization step. This method is supposed to show the benefit of well chosen crossover points. In the following heuristic in each step $k$ randomly chosen crossover settings are tested and the best is used for the final recombination procedure. We call this crossover modification $Xover_{opt}$ and show the results in table 6.5. In these experiments we use the settings of the previous section. $Xover_{opt}$ has been tested on the sphere function, rastrigin, rosenbrock with noise, sharp ridge and the constrained problem 2.40. In these experiments, $k = 10$ randomly generated $\nu$ are tested. $Xover_{opt}$ searches within the convex hull[3] of the parental individuals.

On rastrigin, on rosenbrock with noise and on problem 2.40, no significant improvement can be achieved, although $k = 10$ different recombination settings have been tested. Hence, the influence of the strategy parameter vector $\nu$ is

---

[3] The convex hull for a set of points $X$ is the minimal convex set containing $X$.

**Table 6.5.** Experimental results of $\text{Xover}_{opt}$ on 4 continuous test problems. $\text{Xover}_{opt}$ improves the results on the sphere function and the monotone problem sharp ridge.

|       | sphere | rastrigin | roseNoise | sharp | 2.40 |
|-------|--------|-----------|-----------|-------|------|
| best  | 5.97E-201 | 0.99 | 0.68 | -7.52E+256 | -4964.04 |
| worst | 1.49E-190 | 5.96 | 168.90 | -7.66E+240 | -4483.30 |
| mean  | **5.98E-192** | 3.14 | 18.01 | **-3.03E+255** | -4758.69 |
| dev   | 0 | 1.73 | 38.14 | - | 138.09 |

**Table 6.6.** Analysis of the optimization quality testing $k$ recombination settings on the function sphere. The approximation accuracy improves with increasing $k$, but not linearly.

|       | 2 | 3 | 5 | 10 | 15 | 20 | 30 |
|-------|---|---|---|----|----|----|----|
| best  | 1.31E-168 | 4.91E-177 | 1.45E-187 | 5.97E-201 | 7.88E-206 | 1.14E-207 | 5.87E-216 |
| worst | 1.11E-160 | 8.15E-171 | 2.05E-181 | 1.49E-190 | 3.25E-199 | 4.21E-201 | 2.38E-206 |
| mean  | 5.05E-162 | 5.83E-172 | 1.06E-182 | 5.98E-192 | 1.51E-200 | 2.10E-202 | 1.45E-207 |

too weak. But a weak influence on the fitness is a strong argument that self-adaptation of this parameter must fail. What is the reason for the weakness? Obviously, for the considered multimodal fitness landscapes we can assert the assumption that generating points within the convex hull of the population is not advantageous. The results on the function sphere, see also table 6.6 and sharp ridge are much better. $\text{Xover}_{opt}$ is helpful on ridge functions because the points can concentrate on one edge of the hull due to the monotony of the function. On the sphere function the feature of monotony also helps, because every step into the direction of better fitness is a step into the direction of the optimum. This condition of locality is not given for multimodal functions.

In order to test whether the fitness gain is too sparse for successful self-adaptation, we test a couple of settings for parameter $k$ on the sphere function, see table 6.6 for the results. An improvement can be achieved with every increase of $k$. But the fitness gain is not logarithmically linear depending on $k$, but rather less.

But why is the self-adaptation of $\nu$ on the sphere function not successful? If the points are randomly distributed around the optimum, an *optimal* linear combination of two points exists, but its factors depend on the current situation and not on the problem structure. Hence, the same factors for the linear combination are not optimal in the next generation, the situation may change significantly. This is why the success of inheriting the factors $\nu$ can not be guaranteed and may even be doubtful.

## 6.6  Summary

*(...) Although these results provided valuable insight and have informed many practical implementations, it is worth bearing in mind that they are only*

*considering the disruptive effects of operators. Analysis of the constructive effects of operators in creating new instances of schema H are harder, since these effects depend heavily on the constitution of the current population.*
Eiben about the results of the schema theorem [38], p. 192.

1. The answer whether an improvement of EAs with self-adaptive features of crossover operators can be gathered, is negative. Our experiments show that self-adaptive crossover achieves no *significant* improvements on combinatorial problems like the TSP, on continuous optimization problems, as well as on bit-string problems. Although no significant results can be reported, no deterioration has been observed. But the variants with *changing*, i.e. random or self-adaptive, crossover points are better than the operators with constant points.

2. The results of the crossover point optimization EA shows that on unimodal and monotone functions, optimal crossover points can be found with an *optimization step.*

3. We see two reasons for the *failure* of self-adaptive crossover:

    a) The fitness gain achieved by *optimal* crossover settings is weak for many problems, e.g. for multimodal continuous problems. Our experiments could only discover valuable fitness gain on monotone functions like sphere and ridge. In other words: it is easier to control the estimated distribution functions of mutation than identifying the successive useful crossover points or settings self-adaptively.

    b) Even if the gain of a specific crossover strategy parameter set is advantageous it may not be optimal in the following generation. The same argument holds for the crossover points of PMX and N-point crossover variants. Whenever a successful crossover step has been applied, there seems to be a better crossover point set in the following generation. Hence, a self-adaptive process can not be established.

4. For complex problems we see the challenge to tighten the link between strategy parameter adaptation and fitness gain: choosing individuals from different mating pools or subpopulations with different features may lead to improvements. If individuals are selected from these different sets, their combination at specific crossover points may make sense. But even though, the two mentioned arguments might prevent the success of self-adaptive crossover.

# Part III

# Constraint Handling

# 7 Constraint Handling Heuristics for Evolution Strategies

Whenever the search space is restricted due to constraints of the underlying problem, the EA has to make use of heuristic extensions which are called constraint handling methods. Constraint handling is very relevant to practical applications. A constraint is a restriction on possible value combinations of variables. EAs and in particular ES are used for constrained numerical parameter optimization. The optimum quite often lies on the constraint boundary or even in a vertex of the feasible search space. In such cases the EA frequently suffers from premature convergence because of a low success probability near the constraint boundaries. We prove premature step size reduction for a (1+1)-EA under simplified conditions, analyzing the success rates at the constraint boundary and the expected changes of the step sizes.

First of all, in section 7.1 the numerical NLP-problem is defined. After a short survey of constraint-handling techniques in 7.2, premature fitness stagnation is proven theoretically for simplified conditions and experimentally analyzed in section 7.3. In section 7.4 we present a new constraint-handling method, the death penalty step control evolution strategy (DSES). The DSES uses an adaptive mechanism that controls the reduction of a minimum step size depending on the distance to the infeasible search space. In section 7.5 we introduce a biologically inspired concept of sexual selection and pairing for handling constraints. A number of modifications is necessary to prevent an explosion or the mentioned stagnation of step sizes and to improve the efficiency of the approach. In 7.6 we present the nested angle evolution strategy (NAES). This is a meta-evolutionary approach, in which the angles of the correlated mutation of an inner problem-solving ES are adapted by an outer ES. All techniques are experimentally evaluated on selected test problems and compared with the results of an existing penalty-based constraint-handling method.

## 7.1 The NLP Problem

In general, the constrained continuous nonlinear programming problem is defined as follows: In the $N$-dimensional search space $\mathbb{R}^N$ find a solution $\mathbf{x} = (x_1, x_2, \ldots, x_N)^T$, which minimizes $f(\mathbf{x})$:

O. Kramer: Self-Adaptive Heuristics for Evolutionary Computation, SCI 147, pp. 117–140, 2008.
springerlink.com © Springer-Verlag Berlin Heidelberg 2008

$$f(\mathbf{x}) \rightarrow \min, \ \mathbf{x} \in \mathbb{R}^N \quad \text{with subject to}$$
$$\text{inequalities} \quad g_i(\mathbf{x}) \leq 0, \ i = 1, \ldots, n_1, \ \text{and} \tag{7.1}$$
$$\text{equalities} \quad h_j(\mathbf{x}) = 0, \ j = 1, \ldots, n_2$$

A feasible solution $\mathbf{x}$ satisfies all $n_1$ inequality and $n_2$ equality constraints. Many constraint-handling techniques like penalty functions make use of a constraints violation measurement $G$:

$$G(\mathbf{x}) = \sum_{i=1}^{n_1} \max[0, g_i(\mathbf{x})]^\beta + \sum_{j=1}^{n_2} |h_j(\mathbf{x})|^\gamma \tag{7.2}$$

The parameters $\beta$ and $\gamma$ are usually chosen as one or two. In the following, only inequality constraints are taken into account. Equality constraints can be transformed into inequality constraints as shown in section 7.4.2.

## 7.2   A Short Survey of Constraint-Handling Methods

There exists a variety of constraint-handling techniques for EA. Most of them can be classified into four main types of concepts.

- *Penalty functions* decrease the fitness of infeasible solutions by taking the number of infeasible constraints or the distance to feasibility into account.
- *Repair algorithms* produce feasible solutions from infeasible ones and replace them or use their fitness for evaluation.
- *Decoder functions* map genotypes to phenotypes which are guaranteed to be feasible.
- *Feasibility preserving representations* or *genetic operators* force candidate solutions to be feasible.

The methods are described in more detail in the following.

### 7.2.1   Penalty Functions

Most of the constraint handling methods are based on penalty functions. An early, rather general penalty approach is the SUMT (sequential unconstrained minimization technique) by Fiacco and McCormick [40]. The constrained problem is solved by a sequence of unconstrained optimizations in which the penalty factors are stepwise intensified. In other penalty approaches penalty factors can be defined statically [60] or depending on the number of satisfied constraints [79]. They can dynamically depend on the number of generations of the EA as Joines and Houck propose [66]:

$$\tilde{f}(\mathbf{x}) = f(\mathbf{x}) + (C \cdot t)^\alpha \cdot G(\mathbf{x}) \tag{7.3}$$

The parameter $t$ represents the actual generation, the parameters $C$ and $\alpha$ must be defined by the user. Typical settings are $C = 0.5$, $\alpha = 1$ or 2. Penalties can be adapted according to an external cooling scheme [66] or by adaptive heuristics [11]. In the death penalty approach [75] infeasible solutions are rejected and new solutions are created until enough feasible ones exist.

### 7.2.2  Penalty-Related Methods

Many methods revert to the penalty principle. In the segregated genetic algorithm by Le Riche et al. [118] two penalty functions, a weak and an intense one, are calculated in order to surround the optimum. In the coevolutionary penalty function approach by Coello [24] the penalty factors of the inner EA are adapted by an outer EA. Some methods are based on the assumption that any feasible solution is better than any infeasible [109], [32]. Examples are the metric penalty functions by Hoffmeister and Sprave [57]. Feasible solutions are compared using the objective function while infeasible solutions are compared considering the satisfaction of constraints.

### 7.2.3  Repair Algorithms

Repair algorithms either replace infeasible solutions or only use the repaired solutions for evaluation of their infeasible pendants [25], [12]. Repair algorithms can also be seen as local search methods which reduce the constraint violation. The repair algorithm generates a feasible solution from an infeasible one. In the Baldwinian case, the fitness of the repaired solution replaces the fitness of the original solution. In the Lamarckian case, the feasible solution overwrites the infeasible one. In general defining a repair algorithms can be as complex as solving the problem itself.

### 7.2.4  Decoder Functions

Decoders build up a relationship between the constrained search space and an artificial search space easier to handle [94], [71]. They map a genotype into a feasible phenotype. By this means even quite different genotypes may be mapped onto the same phenotype. Eiben and Smith [38] define decoders as a class of mappings from the genotype space $\mathcal{S}'$ to the feasible regions $\mathcal{F}$ of the solution space $\mathcal{S}$ with the properties:

- Every $z \in \mathcal{S}'$ must map to a single solution $s \in \mathcal{F}$.
- Every solution $s \in \mathcal{F}$ must have at least one representation $s' \in \mathcal{S}'$.
- Every $s \in \mathcal{F}$ must have the same number of representations in $\mathcal{S}'$ (this need not be one).

Michalewicz [94] defined decoder functions for continuous problems.

### 7.2.5  Multiobjective Optimization

Multiobjective optimization techniques are based on the idea of handling each constraint as an objective. Under this assumption many multiobjective optimization methods can be applied. Such approaches were used by Parmee and Purchase [104], Jimenez and Verdegay [65], Coello [23] and Surry et al. [150]. In the behavioral memory-method by Schoenauer and Xanthakis [130] the EA concentrates on minimizing the constraint violation of each constraint in a certain order and optimizing the objective function in the last step.

### 7.2.6  Constraint Preserving Operators and Representations

A further method is to avoid infeasible solutions by special constraints preserving representations and operators. An example is the GENOCOP-algorithm [94] that reduces the problem to convex search spaces and linear constraints. A predator-prey approach to handle constraints is proposed by Paredis [103] using two separate populations. Schoenauer and Michalewicz [129] propose special operators that are designed to search regions in the vicinity of active constraints. A comprehensive overview to constraint handling techniques is given by Coello [25] and also by Michalewicz [94].

### 7.2.7  Recent Developments

Recently, Coello [96] introduced a technique based on a multimembered ES combining a feasibility comparison mechanism with several modifications of the standard ES. At the CEC 2006 special session on constrained real-parameter optimization a number of interesting methods was introduced. The $\epsilon$ constrained differential evolution approach by Takahama [152] combines the usage of an $\epsilon$ for equality constraints with the differential evolution approach. It furthermore uses a gradient-based mutation to find feasible points considering the gradient of infeasible points. Additionally, a scheme of feasible elites is used to improve the search for feasible solutions. The dynamic multi-swarm particle optimizer by Liang and Suganthan [84] makes use of a set of sub-swarms concentrating on different constraints. It is combined with sequential quadratic programming as a local search method. The approach of Mezura-Montes et al. [92] combines differential evolution, different mutation operators to increase the probability of producing better offspring, three selection criteria and a diversity mechanism. Another interesting approach is the population-based method of Sinha et al. [140] using a *parent centric* procedure. It obtained successful results on the CEC test problems.

## 7.3  Premature Step Size Reduction

ES on constrained optimization problems suffer from premature step size reduction in case of active inequality constraints. This results in premature convergence. Broadly speaking, the reason for the premature step size reduction is the fact that the constrained region cuts off the mutative success area. Consequently, the self-adaptation process favors smaller step sizes, whose success area is not cut off. At first, premature step size reduction is analyzed experimentally. Afterwards we prove theoretically that the step sizes tend to reduce at the constraint boundary.

### 7.3.1   Experimental Analysis

The premature step size reduction can be shown experimentally on problem 2.40 for the death penalty method and the dynamical penalty function by Joines and Houck [66], see Table 7.1. Problem 2.40 exhibits a linear objective function and an optimum with five active linear constraints. Each row of Table 7.1 shows the results of a (15,100)-ES after 50 runs. As termination condition fitness stagnation is chosen. If the difference between the fitness value of the best individual of a generation and the best of the following generation is smaller than a $\theta = 10^{-12}$, then the ES terminates as the magnitude of the steps sizes is too small to effect further improvements. Both constraint-handling methods are not able to approximate the optimum of the problem satisfactorily. The standard deviations *dev* show that the algorithms produce rather different results in the various runs.

**Table 7.1.** Experimental results of the death penalty method (DP) and the dynamic penalty function by Joines and Houck (Dyn) on problem 2.40. The parameter *ffc* counts the fitness function calls and *cfc* the constraint function calls. Both constraint-handling techniques are not able to approximate the optimum of the problem satisfactorily. The relatively high standard deviations *dev* show that the algorithms produce unsatisfactorily different results.

|     | best | mean | worst | dev | ffc | cfc |
|-----|------|------|-------|-----|-----|-----|
| DP  | -4948.079 | -4772.338 | -4609.985 | 65.2 | 50624 | 96817 |
| Dyn | -4780.554 | -4559.129 | -4358.446 | 85.0 | 31878 | 31878 |

### 7.3.2   Theoretical Analysis

We consider the two dimensional case of a linear objective function and one linear constraint with the angle $\beta$ between the latter and the contour lines of the fitness function. Figure 7.1 shows a typical situation of individual $\mathbf{x}$ in the neighborhood of the constraint boundary. It shows the success rate situations of individual $\mathbf{x}$ with distance $d$ to the constraint boundary for the three cases: 1. $\sigma < d$, 2. $\sigma > d$, $\sigma < s$ and 3. $\sigma > d$, $\sigma > s$.

We analyze the behavior of a (1+1)-EA with adaptive step sizes modeled[1] by the Markovian process $(X_t, \sigma_t)_{t \geq 0}$ generated by

$$X_{t+1} = \begin{cases} X_t + \sigma_t Z_t & \text{if } f(X_t + \sigma_t Z_t) < f(X_t) \wedge g(X_t + \sigma_t Z_t) = 0 \\ X_t & \text{otherwise} \end{cases} \quad (7.4)$$

---

[1] As Rudolph [126] states, this EA does not exactly match a (1+1)-EA with self-adaptive step size control, but it can be transferred to a broader class of EAs.

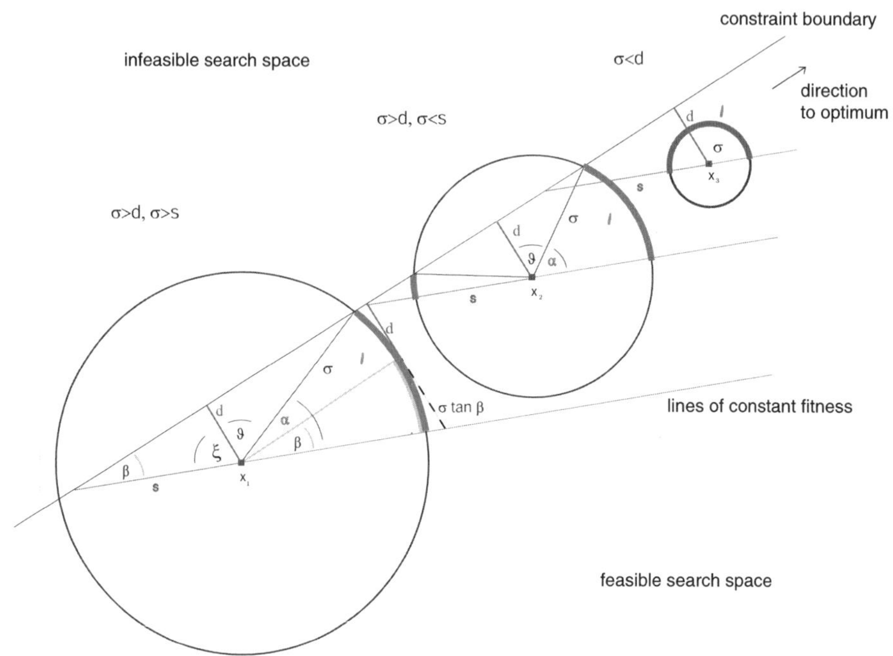


**Fig. 7.1.** Success rates at the boundary of the feasible search space. Three cases have to be considered, i.e. 1. $\sigma < d$, 2. $\sigma > d$, $\sigma < s$ and 3. $\sigma > d$, $\sigma > s$. The bold circular arcs are the regions where successful mutations are produced.

and

$$
\sigma_{t+1} = \begin{cases} \gamma \sigma_t & \text{if } f(X_t + \sigma_t Z_t) < f(X_t) \wedge g(X_t + \sigma_t Z_t) = 0 \\ \gamma^{-1}\sigma_t & \text{otherwise} \end{cases}
\tag{7.5}
$$

with step size $\sigma_t$ and mutation parameter $\gamma > 1$. The function $g$ measures the constraint violation. Each random vector $Z_t, t \geq 0$ is independent and identically distributed in the following way: We assume that mutations $\sigma_t Z_t$ are produced on the edge of the circle around $X_t$ with radius $\sigma_t$. When a successful mutation is produced, the step length $\sigma_t$ is increased and decreased otherwise. We are interested in the development of the step size $\sigma_t$ and the distance $d_t$ to the constraint boundary. For the sake of better readability we write $\sigma$ instead of $\sigma_t$ and $d$ instead of $d_t$ where possible. In the following lemma 7.1 we analyze the success probabilities for the three cases.

**Lemma 7.1.** *Let $(p_s)$ be the success probability for individual $X_t$ of the (1+1)-EA, with step size $\sigma$ and distance $d$ to the constraint boundary. Then it holds $(p_s)_{\sigma<d} = 1/2$ and $(p_s)_{\sigma>d} < 1/2$. For $d/\sigma \to 0$ it holds $(p_s)_{\sigma>d} \to \beta/(2\pi)$.*

*Proof.* The analysis of the probabilities for changing through the states is based on a success rates analysis of the circle model for the three mentioned cases, see figure 7.1. In our model the success rate $p_s$ is the relation between the length of the circular arc $l = 2\pi r \alpha / (2\pi)$ and the circumference $2\pi r$:

$$p_s = \alpha/(2\pi). \tag{7.6}$$

The EA starts its run in the feasible part of the search space, $\sigma < d$. As the fitness function is linear and the constraint boundary does not cut off the circle, the angle over the circular arc with successful mutations is $\alpha = \pi$. Hence,

$$\boxed{(p_s)_{\sigma<d} = 1/2.} \tag{7.7}$$

The step sizes are increased with probability $1/2$ and decreased with probability $1/2$. The constrained case is much more important. For $\sigma > d$ we have to distinguish two cases:

1. $\sigma < s$, i.e. the circle cuts its contour line within the feasible region. Distance $s$ is the segment of the contour line between $x$ and the constraint boundary with $s = (d/\sin\beta)$. The success area on the edge of the circle is the circular arc over $\alpha$ and the circular arc in the opposite direction. The success probability $(p_s)_{\sigma>d,\sigma<s}$ can also be expressed with the help of $\vartheta$

$$(p_s)_{\sigma>d,\sigma<s} = 1/2 - \vartheta/\pi \tag{7.8}$$

The triangle with a right angle yields $cos(\vartheta) = d/\sigma$ and the success probability becomes

$$(p_s)_{\sigma>d,\sigma<s} = 1/2 - \arccos(d/\sigma)/\pi \tag{7.9}$$

As $0 < (d/\sigma) < 1$, it holds $0 < \arccos(d/\sigma) < \pi/2$ and the desired property

$$\boxed{(p_s)_{\sigma>d,\sigma<s} < 1/2.} \tag{7.10}$$

The success probability $(p_s)_{\sigma>d,\sigma<s}$ is comparatively small. Analyzing the asymptotic behavior we get

$$(p_s)_{\sigma>d,\sigma<s} \to 0 \text{ for } d/\sigma \to 0, \tag{7.11}$$

i.e. for small distances $d \to 0$ or huge step sizes $\sigma \to \infty$ the success probability becomes 0. But one have to keep in mind that a small $d/\sigma$ implies a small angle $\beta$ as $\sigma < s$. It is obvious that $(p_s)_{\sigma>d,\sigma<s} \geq (p_s)_{\sigma>d,\sigma>s}$. Hence the analysis of the case $\sigma > s$ is more interesting as it considers $\beta$.

2. $\sigma > s$, i.e. the circle cuts its contour line beyond the constraint boundary. The angle $\alpha$ is $\pi - (\xi + \vartheta)$. In the triangle of $\beta$ and $\xi$ it holds $\xi = \pi/2 - \beta$, we get $\alpha = \beta + \pi/2 - \arccos(d/\sigma)$. Hence, the success probability is

$$(p_s)_{\sigma>d,\sigma>s} = \alpha/(2\pi) = \beta/(2\pi) + (\pi/2 - \arccos(d/\sigma))/(2\pi) \qquad (7.12)$$

Again, for $0 < d/\sigma < 1$, it holds $0 < \arccos(d/\sigma) < \pi/2$ and as we postulated $\beta = \pi/2$, we get

$$\boxed{(p_s)_{\sigma>d,\sigma>s} < 1/2.} \qquad (7.13)$$

As $(p_s)_{\sigma>d,\sigma<s} \geq (p_s)_{\sigma>d,\sigma>s}$ for a small distance $d$ or a big step size $\sigma$, we can assert the asymptotic behavior

$$(p_s)_{\sigma>d} \to \beta/(2\pi) \text{ for } d/\sigma \to 0 \qquad (7.14)$$

Furthermore, $(p_s)_{\sigma>d}$ decreases for smaller $\beta$. $\qquad\qquad\square$

*Example 7.2.* We give an example to illustrate the success rate situation. We assume that $\beta = \pi/4$ and $d/\sigma = 1/2$. As $2d > (d/\sin\beta)$, we have to consider the case $\sigma > s$. The success rate becomes $(p_s)_{\sigma>d,\sigma>s} = 5/24$. As the probability for a small step size $(p_s)_{\sigma<d} = 1/2$ is much bigger, it will in particular be preferred by a self-adaptive mechanism.

We continue to prove the step size reduction for our (1+1)-EA. As the optimum lies at the constraint boundary, the EA will *encounter* the latter, $\sigma > d$. Theorem 7.3 describes the behavior of the (1+1)-EA at the constraint boundary by stating the expected mutation change $E(\gamma)$ in each iteration, lemma 7.1 provides a quantitative analysis stating the success rate $(p_s)_{\sigma>d}$, in the following denoted as $p$.

**Theorem 7.3 (Premature Step Size Reduction).** *Let $f$ be a linear fitness function and $g$ a linear constraint boundary with angle $\beta < \pi/2$ between $f$ and $g$. In the vicinity of the constraint boundary $\sigma > d$, the above modeled (1+1)-EA with $\gamma > 1 + (\tan\beta)$ reduces its step size $\sigma$ in each iteration by $E(\gamma) = \gamma^{\frac{1}{(1-p)} - \frac{1}{p}} < 1$ with $p < 1/2$.*

*Proof.* We distinguish two states $\Gamma_1$ and $\Gamma_2$ for an individual $X_t$ in the neighborhood of the constraint boundary, $\sigma > d$.

1. $\Gamma_1$ denotes the situation that a successful mutation was produced, i.e. $f(X_t + \sigma_t Z_t) < f(X_t) \wedge g(X_t + \sigma_t Z_t) = 0$.
2. $\Gamma_2$ denotes the failure $f(X_t + \sigma_t Z_t) > f(X_t) \vee g(X_t + \sigma_t Z_t) > 0$. Lemma 7.1 shows that the probability for a success is small, at most smaller than $1/2$ and converges to $\beta/(2\pi)$ for $d/\sigma \to 0$.

The state transitions between the two states $\Gamma_1$ and $\Gamma_2$ are analyzed in the following.

- $\Gamma_1 \rightarrow \Gamma_1$. The probability for a success is $p$, which is relatively low, see lemma 7.1. A success results in a step size increase $\sigma_{t+1} = \gamma\sigma_t$. A successful mutation may lie arbitrarily close to the constraint boundary $(d/\sigma \rightarrow 0)$ and therefore decrease the success probability $p$ rapidly. But a success may also lead to an increase of the distance to the constraint boundary, at most by $\sigma \tan \beta$. The constrained case is not left, if the step size increase is higher than the distance increase to guarantee $d/\sigma < 1$:

$$\frac{d + \sigma \tan \beta}{\sigma \gamma} < 1 \equiv \frac{d}{\sigma} + \tan \beta < \gamma \tag{7.15}$$

As $d/\sigma < 1$ it must hold $\gamma > 1 + (\tan \beta)$ to fulfill the above condition for the proof of step size reduction. So, the probability for staying in state $\Gamma_1$ is

$$P(\Gamma_1 \rightarrow \Gamma_1 | \Gamma_1) = p, \quad \sigma_{t+1} = \gamma\sigma_t. \tag{7.16}$$

- $\Gamma_2 \rightarrow \Gamma_1$: The probability for a success if the last step was a failure is again $p$, the step size is increased and the constrained case is not left for $\gamma > 1 + \tan \beta$.

$$P(\Gamma_2 \rightarrow \Gamma_1 | \Gamma_2) = p, \quad \sigma_{t+1} = \gamma\sigma_t. \tag{7.17}$$

- $\Gamma_1 \rightarrow \Gamma_2$: The probability for a failure is $1 - p$. It results in a step decrease $\sigma_{t+1} = \gamma^{-1}\sigma_t$. If the step decrease leads to $d/\sigma_{t+1} > 1$, the constraint boundary is left. In this case step size decrease and increase occur with the same probability and the expected change of $\sigma$ becomes $E(\gamma) = \gamma \cdot \gamma^{-1} = 1$. But the constraint boundary will be reached again within the following steps. Hence, we summarize

$$P(\Gamma_1 \rightarrow \Gamma_2 | \Gamma_1) = 1 - p, \quad \sigma_{t+1} = \gamma^{-1}\sigma_t. \tag{7.18}$$

- $\Gamma_2 \rightarrow \Gamma_2$: Similar to transition $\Gamma_1 \rightarrow \Gamma_2$ the probability to stay in the state $\Gamma_2$ is the probability $1 - p$ for a failure.

$$P(\Gamma_2 \rightarrow \Gamma_2 | \Gamma_2) = 1 - p, \quad \sigma_{t+1} = \gamma^{-1}\sigma_t. \tag{7.19}$$

This yields the following state transition probability matrix $\mathbf{T}$ for states $\Gamma_1$ and $\Gamma_2$:

$$\mathbf{T} = \begin{pmatrix} p & (1-p) \\ p & (1-p) \end{pmatrix} \tag{7.20}$$

It is worth to mention that in the case of a success with an overwhelming probability of $p' = (1 - \beta/(2\pi)) \rightarrow 1$ for $\beta \rightarrow 0$, the distance $d$ to the constraint

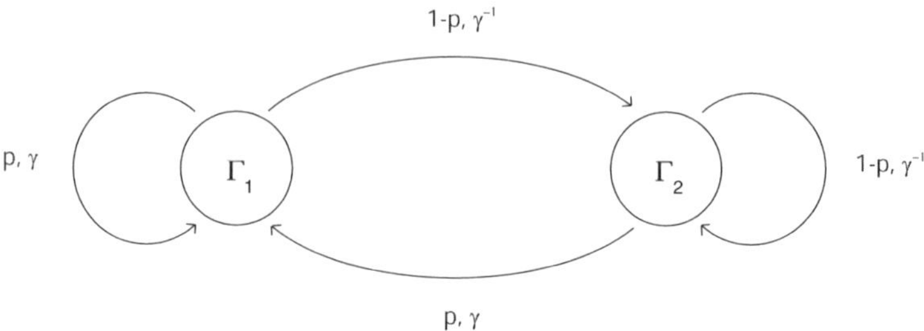


**Fig. 7.2.** The state transitions of $\Gamma_1$ and $\Gamma_2$, its probabilities at the constraint boundary and the influence on the step size $\sigma$

boundary is decreased. This condition also contributes to an iterative reduction of the distance to the constraint boundary before reaching the optimum. From the probability of each state transition and the step size change, an expected step size change can be derived. Figure 7.2 shows the probabilities for the state transitions of $\Gamma_1$ and $\Gamma_2$, together with the change of step size $\sigma$ for each transition.

In each state, there are only two possibilities: staying in the same state or leaving the state. Hence, the state transitions are geometrically distributed. We are able to determine the *expected* change of the step sizes. The probability to *leave* state $\Gamma_1$ is $1-p$, so the expected number of iterations to stay is $1/(1-p)$. Each iteration the step size is increased by $\gamma$. When leaving to state $\Gamma_2$ the step size is decreased by $\gamma^{-1}$. The probability to *leave* state $\Gamma_2$ is $p$, so the expected number of iterations to stay is $1/p$ with a step decrease by $\gamma^{-1}$. Returning to state $\Gamma_1$ leads to a step increase by $\gamma$. From these considerations we can now determine the expected development of $\gamma$:

$$E(\gamma) = \gamma^{\frac{1}{1-p}} \cdot \gamma^{-1} \cdot \gamma^{-\frac{1}{p}} \cdot \gamma = \gamma^{\frac{1}{1-p} - \frac{1}{p}} \tag{7.21}$$

Lemma 7.1 has proven that $p < 1/2$, so $\frac{1}{1-p} - \frac{1}{p} < 0$ and

$$\boxed{E(\gamma) = \gamma^{\frac{1}{1-p} - \frac{1}{p}} < 1} \tag{7.22}$$

The expected change of step size $\sigma$ is $E(\gamma) < 1$ in each generation, so the steps are decreasing at the constraint boundary. For small success probabilities, see lemma 7.1, the decrease becomes quite high.    $\square$

We proved the step size reduction theoretically for the (1+1)-EA and experimentally for a (15,100)-ES in section 7.3.1. In the next sections we present three heuristics to overcome the problem of premature step size reduction.

## 7.4  The Death Penalty Step Control Approach

In the following we introduce the death penalty step control evolution strategy (DSES). It is based on a least step size, which has to be controlled, in order to allow convergence to the optimum.

### 7.4.1  Basic Minimum Step Size Reduction Mechanism

As mentioned in section 7.3 the death penalty method suffers from premature step size reduction. The DSES is based on death penalty, i.e. rejection of infeasible solutions. For the initialization feasible starting points are required. The concept of the approach is a minimum step size $\epsilon$, a lower bound on the step sizes $\sigma$ that prevents the evolutionary process from premature step size reduction. But it also prevents the optimization process from unlimited convergence to the optimum when reaching the range of $\epsilon$. Consequently, a control mechanism is introduced with the task of reducing $\epsilon$ when approximating the optimum. Intuitively, the reduction process depends on the number of infeasible mutations produced when reaching the area of the optimum at the boundary of the feasible search space. Consider the situation presented in Figure 7.3. For the sake of better understanding we do as if all mutations fall within a $\sigma$-circle around the parental individual. On the left (a), the individual $\mathbf{x}$ has come quite close to the optimum at a vertex of the feasible search space. Further approximation (b) with the same minimum step size means an increase of infeasible mutations which are counted with the parameter $z$. The reduction process of $\epsilon$ depends

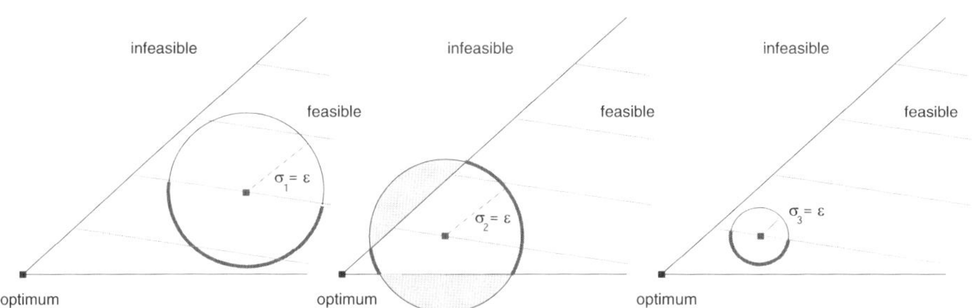


**Fig. 7.3.** Concept of the DSES. The optimum lies in a vertex of the feasible search space. For the sake of better understanding we do as if all mutations fall into a $\sigma$-circle around the parental individual instead of using a normal distribution with standard deviation $\sigma$. (a) The individual $\mathbf{x}$ approximates the optimum. (b) A further approximation is possible because of $\epsilon$, until the marked region of success becomes considerably small and many mutations fall into the infeasible region. (c) When the number of infeasible trials exceeds the parameter $\varpi$ the minimum step size $\epsilon$ is reduced and a further approximation of the optimum becomes possible.

```
1     Start
2         t:=0;
3         Initialize parental population P(t)
4         Repeat
5           For k=1 To λ Do
6              z = 0;
7              Repeat
8                 Choose ρ parents from P(t)
9                 recombination_step_sizes;
10                recombination_objective_variables;
11                mutation_step_sizes;
12                If (z mod mod) == 0  Then
13                   ε = ε· melt;
14                For j=1 To N Do
15                  If σ_j^k < ε Then
16                     σ_j^k = ε
17                Next
18                mutation_objective_variables;
19                fitness of a_k := f(x);
20                z:=z+1;
21              Until a_k feasible
22              Add a_k to offspring population O(t);
23           Next
24           Select parental population P(t+1) from O(t);
25           t := t+1;
26        Until termination condition
27     End
```

**Fig. 7.4.** Pseudocode of the DSES

on the number $z$ of rejected infeasible solutions: Every $\varpi$ infeasible trials, $\epsilon$ is reduced by a factor $0 < \vartheta < 1$ according to the equation:

$$\epsilon' := \epsilon \cdot \vartheta \tag{7.23}$$

The DSES is denoted by $[\varpi; \vartheta]$-DSES. Figure 7.4 shows the pseudocode of the DSES.

## 7.4.2  Experimental Analysis

We summarize the results of our experimental analysis on the various constrained problems [72], [75]. Note that we translate equality constraints $h_j(\mathbf{x}) = 0$, $j = 1,\ldots,n_2$ into inequality constraints $g_i(\mathbf{x}) \leq \epsilon$, $i = n_1 + 1,\ldots,n_2$ and $g_i(\mathbf{x}) \leq -\epsilon$, $i = n_2 + 1,\ldots,2n_2$ and $\epsilon = 0.0001$. For all ES we used $n_\sigma = N$ step sizes ($N$ is the dimension of the problem) and the ES mutation parameter recommendations for $\tau_0$ and $\tau_1$, see the experimental settings. We chose the initial step sizes $\sigma_i = \frac{|x^{(0)} - x^*|}{N}$ with $1 \leq i \leq N$, starting point $x^{(0)}$ and optimum $x^*$ according to the recommendation of Schwefel [135]. As termination condition we chose premature mutation strength reduction, i.e. the ES is terminated if the difference between the best individuals of two successive generations is smaller than $\theta = 10^{-12}$. All DP and DSES tests are based on a (15/2,100)-ES with intermediary recombination.

---

**Experimental settings**

| | |
|---|---|
| Population model | (15,100) |
| Mutation type | standard, $n_\sigma = N$, $\tau_0 = (\sqrt{2n})^{-1}$ and $\tau_1 = (\sqrt{2\sqrt{n}})^{-1}$ |
| Crossover type | intermediate, $\rho = 2$ |
| Selection type | comma |
| Initialization | $\sigma_i = \frac{|x^{(0)} - x^*|}{N}$ |
| Termination | fitness stagnation, $\theta = 10^{-12}$ |
| Runs | 25 |

---

The measured parameters are the usual ones concerning the fitness of the best individual in the last generation of the various runs (*best*, *avg*, *worst* and *dev*). Parameter $\sigma$ is the mean of the mutation strength in the last generation and therefore an indicator for premature stagnation. *Ffc* counts the number of fitness function calls, *cfc* the number of constraint function calls.

## Parameter Analysis of the DSES

Table 7.2 shows the analysis of the DSES parameter settings. Above, the analysis of the DSES on problem g01 reveals a significant improvement in comparison to DP. The [400;0.5]-, the [400;0.3]- and the [400;0.1]-DSES were able to approximate the optimum with arbitrary accuracy. The $\epsilon$-reduction was performed too fast for the [100,$\vartheta$]-DSES with $\vartheta = 0.1, 0.3, 0.5$ in every run.

We analyze the behavior of the DSES on problem g07. In comparison to the method death penalty a significant improvement of accuracy can be observed. The slower $\epsilon$ is decreased the more the quality of the results can be improved. But we have to admit that *ffc* and in particular *cfc* explode. Hence, we can only recommend the DSES when *ffc* and *cfc* are not too expensive. The experiments show that death penalty completely fails concerning the quality of the results.

## The DSES on All Functions

The experimental results of the DSES on all considered problems are summarized in table 7.3. On Schwefel's problems 2.40 and 2.41 the DSES is able to approximate the optimum with most of the tested settings. DP fails on these problems and suffers from premature mutation strength reduction. Problem TR2 is hard to tackle. But the results of a [15; 0.5]-DSES on TR2 are better than the results of DP. As expected the number of *ffc* and *cfc* explode, i.e. they are

**Table 7.2.** Above: Experimental analysis of a $(15/2, 100)$-DSES with intermediary recombination and an initial step size of $\sigma_i = 0.1$ on problem g01. The DSES is able to approximate the optimum with arbitrary accuracy while DP fails on g01. Below: A $(15/2, 100)$-DSES with intermediary recombination on problem g07. We observe significant better results in comparison to DP and achieve the best results with the slowest $\epsilon$-reduction. On the other hand the slow reduction causes an explosion of constraint function calls. The constraint violation of each best solution is 0.

| DSES | best | avg | worst dev | $\sigma$ | ffc | cfc |
|---|---|---|---|---|---|---|
| *g01* | | | | | | |
| [400; 0.5] | −15.0 | −14.9999999991 | −14.99999999 1.7E-9 | 2.0E-10 | 93 224 | 3 174 166 |
| [400; 0.3] | −15.0 | −14.9999999988 | −14.99999998 2.3E-9 | 2.1E-10 | 75 484 | 2 035 043 |
| [400; 0.1] | −15.0 | −14.9999999976 | −14.99999995 8.9E-9 | 3.7E-10 | 55 992 | 1 136 101 |
| [100; 0.5] | −15.0 | −14.9199999995 | −12.99999999 0.399 | 6.8E-11 | 51 764 | 548 682 |
| [100; 0.3] | −15.0 | −14.8181247506 | −12.45311878 0.634 | 1.3E-10 | 40 372 | 312 837 |
| [100; 0.1] | −15.0 | −14.7537488912 | −12.84372234 0.681 | 2.6E-10 | 37 716 | 260 461 |
| DP | −15.0 | −14.1734734986 | −11.58255662 1.106 | 5.1E-11 | 33 067 | 110 120 |
| *g07* | | | | | | |
| [70; 0.7] | 24.306237739205 | 24.3067230937 | 24.30819910 4.5E-4 | 3.8E-9 | 1655956 | 11159837 |
| [70; 0.5] | 24.306330953891 | 24.3074746341 | 24.31034458 9.5E-4 | 9.1E-10 | 936436 | 6287404 |
| [70; 0.3] | 24.306433546473 | 24.3095029336 | 24.32761195 0.004 | 1.7E-9 | 577736 | 3850091 |
| [40; 0.7] | 24.308209926624 | 24.3350151850 | 24.37713109 0.019 | 6.3E-11 | 68436 | 401509 |
| [40; 0.5] | 24.315814462518 | 24.3570108927 | 24.47858422 0.039 | 8.3E-11 | 47996 | 263890 |
| [40; 0.3] | 24.337966344507 | 24.4005297646 | 24.57785438 0.055 | 2.1E-10 | 37084 | 189393 |
| DP | 24.449127539670 | 26.3375776765 | 30.93483255 1.472 | 1.2E-11 | 30835 | 87884 |

100 times higher. The behavior of various $[\varpi; \vartheta]$-settings on problem g01 have already been described in section 7.4.2. On g02 the experiments show that also a slow reduction of $\epsilon$ can only improve the quality of the results slightly. It has to be paid with a high number of fitness and constraint function calls. No significant improvement can be achieved on problem g02, although we pay with inefficiency. For the DSES it is not difficult to obtain arbitrary convergence to the optimum of problem g04. On problem g06 all tested DSES variants achieved promising results. But the standard DP has to be recommended as the *cfc* values are lower in comparison. The behavior of the DSES on g07 has already been described. Like on g06 the DSES and DP are able to approximate the optimum of g08 very well. But here, the method DP exhibits no significant performance advantages. Only the slowest decrease of $\epsilon$ on problem g09 enables the DSES to approximate the optimum better than DP and better than *faster* DSES variants. Again, this improvement has to be paid with higher *ffc* and *cfc*. While problem g11 can be approximated sufficiently, the results of the DSES on problems g12 and g16 are exceeding good. As already stated, the method DP performs well on problem g24 while the DSES performs similarly on this problem.

We summarize: the DSES is based on the reduction of a minimum step size which prevents premature step size reduction. It is able to approximate most constrained problems of the *g-test suite*. The DSES is efficient concerning fitness function calls, but inefficient concerning constraint function calls.

**Table 7.3.** Summary of the experimental results of the DSES on the considered problems. The heuristic is able to approximate the optima of most of the problems.

| DSES | best | avg | worst | dev |
|---|---|---|---|---|
| 2.40 $[100; 0.7]$ | $-4999.9999999999$ | $-4999.9999999995$ | $-4999.9999999960$ | 7.99E-10 |
| 2.41 $[75; 0.7]$ | $-17857.1428571428$ | $-17857.1428571425$ | $-17857.1428571404$ | 4.71E-10 |
| TR2 $[15; 0.3]$ | $2.0000000008$ | $2.0000042774$ | $2.0000539455$ | 8.45E-6 |
| g01 $[400; 0.5]$ | $-14.9999999998$ | $-14.9999999991$ | $-14.9999999908$ | 1.78E-9 |
| g02 $[15; 0.3]$ | $-0.8036187549$ | $-0.7658619287$ | $-0.6999587065$ | 0.029 |
| g04 $[25; 0.7]$ | $-30665.5386717833$ | $-30665.5386717831$ | $-30665.5386717826$ | 1.60E-10 |
| g06 $[10; 0.3]$ | $-6961.8138755801$ | $-6961.8138755801$ | $-6961.8138755800$ | 1.90E-11 |
| g07 $[70; 0.7]$ | $24.3062377392$ | $24.3067230937$ | $24.3081991031$ | 4.57E-4 |
| g08 $[2; 0.9]$ | $-0.0958250414$ | $-0.0958250414$ | $-0.0958250414$ | 9.06E-17 |
| g09 $[18; 0.7]$ | $680.6301304921$ | $680.6308434198$ | $680.6322597787$ | 6.59E-4 |
| g11 $[10; 0.5]$ | $0.7499000007$ | $0.7499008144$ | $0.7499035419$ | 1.02E-6 |
| g12 $[200; 0.5]$ | $-0.9999999999$ | $-0.9999999999$ | $-0.9999999999$ | 2.05E-12 |
| g16 $[100; 0.5]$ | $-1.9051552585$ | $-1.9051552584$ | $-1.9051552580$ | 9.12E-11 |
| g24 $[15; 0.3]$ | $-5.5080132715$ | $-5.5080132715$ | $-5.5080132714$ | 2.36E-11 |

## 7.5  Constraint-Handling with Two Sexes

The idea of the concept called two sexes evolution strategy (TSES) is to handle the objective function and the constraint functions as separate objectives. The TSES heuristic is introduced in this section.

### 7.5.1  Biologically Inspired Constraint-Handling

The TSES works as follows. Every individual of the TSES is assigned to a new feature called its sex. Similar to nature, individuals with different sexes are selected according to different objectives. Individuals with sex $o$ are selected by the objective function. Individuals with sex $c$ are selected by the fulfillment of constraints. The intermediary recombination operator plays a key role. Recombination is only allowed between parents of different sex. The treatment of objective function and constraints as separate objectives sounds similar to the multiobjective optimization approaches for constraint-handling. But instead of a multiobjective optimization method, a biologically inspired concept of pairing two sexes is introduced. Consider the situation presented in Figure 7.5. Again, the optimum lies at the boundaries of the feasible search space. The optimum of the unconstrained objective function lies beyond the boundary in the infeasible search space. In the so-called $(\mu_o + \mu_c, \lambda_o + \lambda_c)$-TSES $\mu_o$ parents are selected out of $\lambda_o$ individuals with sex $o$, whereas $\mu_c$ parents are selected out of $\lambda_c$ offspring individuals of the previous generation with sex $c$. As the individuals with sex $o$ are selected according to the objective function, they tend to lie finally in the infeasible search space (pink dots) whereas the $c$-individuals are selected by the fulfillment of all constraints and mostly lie in the feasible search space (green

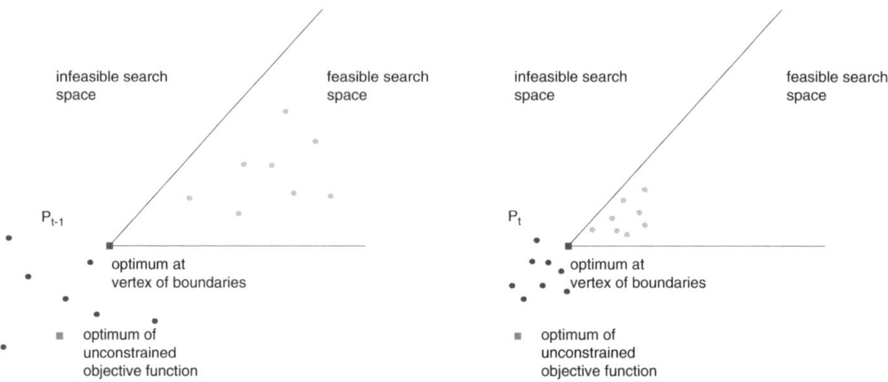


**Fig. 7.5.** Concept of the TSES. The figure shows the effect of intermediary recombination. Individuals can reach a constrained optimum from both sides of the boundaries. Left: The individuals with sex $o$ (pink dots) enter the infeasible region. Right: After intermediary recombination all individuals get closer to the optimum on a vertex of the feasible region.

dots). The measurement $G$ for the fulfillment of constraints has already been defined in equation 7.2. By means of intermediary recombination, all individuals get closer to the optimum of the problem, but still are found on opposite sides of the boundaries between the feasible and the infeasible search space. For the initialization feasible starting points are not required.

### 7.5.2   Modifications of the Basic Two Sexes Evolution Strategy

Several modifications of the basic concept of the TSES are necessary until the algorithm provides successful results. The usual self-adaptation process effects an explosion of the mean step sizes: The invasion far into the feasible search space is rewarded with high fitness values for the $c$-individuals. Approaching the unconstrained optimum of the objective function is rewarded for the $o$-individuals. Modifications of the population ratios of the TSES aim at reducing the diversity in the population to avoid the over-adaptation of the step sizes. Several experiments with different sex ratios and selection operators on problem 2.40 lead to the following heuristic modifications:

- sex ratio and birth surplus with settings around $\lambda_o \approx 15$ and $\lambda_c \approx 85$, Kramer & Schwefel [75] recommend $(8 + 8, 13 + 87)$ ,
- two-step selection operator for the sex $c$, according to the metric penalty function by Hoffmeister and Sprave [57]: Firstly, selection by fulfillment of constraints, secondly, if enough feasible solutions exist, selection by objective function,
- introduction of a finite life span $1 < \kappa < \infty$, see section 2.2.2, for individuals with the sex $c$.

| | |
|---|---|
| 1 | **Start** |
| 2 | t:=0; |
| 3 | Initialize parental population $\mathcal{P}_o(t)$ with sex $o$; |
| 4 | Initialize parental population $\mathcal{P}_c(t)$ with sex $c$; |
| 5 | **Repeat** |
| 6 | **For** k=1 **To** $\lambda_c + \lambda_o$ **Do** |
| 7 | Choose one parent from $\mathcal{P}_o(t)$ and one from $\mathcal{P}_c(t)$ |
| 8 | recombination_step_sizes; |
| 9 | recombination_objective_variables; |
| 10 | mutation_step_sizes; |
| 11 | mutation_objective_variables; |
| 12 | fitness of $a_k := f(\mathbf{x}_k)$; |
| 13 | **If** $k \leq \lambda_o$ **Then** |
| 14 | $\text{sex}(a_k) = o$; |
| 15 | **Else** |
| 16 | $\text{sex}(a_k) = c$; |
| 17 | Add $a_k$ to offspring population $\mathcal{O}(t)$; |
| 18 | **Next** |
| 19 | Select parental population $\mathcal{P}_o(t+1)$ |
| 20 | from $\mathcal{O}(t)$ considering the fitness function; |
| 21 | Select parental population $\mathcal{P}_c(t+1)$ |
| 22 | from $\mathcal{O}(t)$ considering the constraint violation; |
| 23 | $t := t + 1$; |
| 24 | **Until** termination condition |
| 25 | **End** |

**Fig. 7.6.** Pseudocode of the TSES

These modifications lead to promising results on the test functions, see next section 7.5.3. The sex ratio with a majority of $\lambda_c \approx 85$ $c$-individuals and only $\lambda_o \approx 15$ $o$-individuals show that the diversity within the $o$-individuals may not exceed a certain level. Otherwise, the population would be able to reach the region of the unconstrained optimum resulting in an explosion of mean step sizes. The survival possibility for the most successful individuals over up to $\kappa$ reproduction cycles emphasizes the important role of the $c$-individuals. Figure 7.6 shows the pseudocode of the TSES.

### 7.5.3   Experimental Analysis

An experimental analysis shows the properties of the TSES under various conditions.

**Parameter Analysis of the TSES**

First, we show the runs of the TSES under various parameter settings for $\mu_o, \mu_c, \lambda_o, \lambda_c$ and $\kappa$ [72], [75]. Table 7.4, upper part, shows the outcome of the

**Table 7.4.** Above: Experimental analysis of the TSES on problem g01. The standard DP method and the (8+8,13+87)-TSES show bad results, an increase of population sizes is necessary. All other experiments were more successful, the optimum could be reached in almost every run. Below: the TSES on problem g12. The TSES approximates the optimum $-1.0$ in most of the experiments with sufficient accuracy. In some runs of the $(8 + 8, 13 + 87)$-TSES with fixed starting points a premature termination was observed.

| TSES | $\kappa$ | best | avg | worst | dev | $\sigma$ | ffc/cfc |
|---|---|---|---|---|---|---|---|
| *g01* | | | | | | | |
| $(8 + 8, 13 + 87)$ | 200 | $-15.0$ | $-14.8008668542$ | $-12.46168543$ | 5.4E-2 | 1.4E-11 | 360 599 |
| $(20 + 20, 25 + 200)$ | 50 | $-15.0$ | $-15.0$ | $-15.0$ | | 9.1E-15 | 2.7E-14 243 920 |
| $(20 + 20, 25 + 200)^*$ | 50 | $-15.0$ | $-14.8958084010$ | $-12.44021510$ | 4.0E-2 | 2.9E-14 | 277 823 |
| $(40 + 40, 50 + 400)$ | 50 | $-15.0$ | $-15.0$ | $-15.0$ | | 5.3E-15 | 1.8E-14 411 732 |
| $(40 + 40, 50 + 400)^*$ | 50 | $-15.0$ | $-15.0$ | $-15.0$ | | 5.5E-15 | 1.7E-14 468 580 |
| $(15, 100)$-DP | 0 | $-15.0$ | $-14.1734734986$ | $-11.58255662$ | 1.106 | 5.1E-11 | 143 187 |
| *g12* | | | | | | | |
| $(8 + 8, 15 + 85)$ | 50 | $-1.0$ | $-1.0$ | $-1.0$ | 0.0 | 5.5E-8 | 12 305 |
| $(8 + 8, 15 + 85)^*$ | 50 | $-1.0$ | $-1.0$ | $-1.0$ | 0.0 | 5.9E-8 | 11 722 |
| $(8 + 8, 13 + 87)$ | 50 | $-1.0$ | $-0.9999437499$ | $-0.994374999$ | 5.6E-5 | 1.7E-7 | 14 534 |
| $(8 + 8, 13 + 87)^*$ | 50 | $-1.0$ | $-1.0$ | $-1.0$ | 0.0 | 6.9E-8 | 12 720 |
| $(8 + 8, 10 + 90)$ | 50 | $-1.0$ | $-1.0$ | $-1.0$ | | 7.4E-16 | 1.5E-6 18 938 |
| $(8 + 8, 10 + 90)^*$ | 50 | $-1.0$ | $-1.0$ | $-1.0$ | | 1.6E-16 | 1.4E-6 17 237 |
| $(15, 100)$-DP | 0 | $-1.0$ | $-1.0$ | $-1.0$ | | 2.9E-13 | 5.7E-5 20 318 |

experiments on problem g01. Problem g01 exhibits a quadratic objective function and nine linear inequality constraints. The experiments marked with a star make use of randomized starting individuals. The other tests use a fixed initialization. The (8+8,13+87)-TSES was not at all able to approximate the optimum. But a modification of the sex ratios could change the situation. Both, a (20+20,25+200)- and a (40+40,50+400)-TSES, were able to find the optimum with arbitrary accuracy in almost every run. Only the (20+20,25+200)-TSES with randomized starting point suffered from premature step size reduction before reaching the optimum. In comparison to death penalty a significant improvement could be observed.

**Experimental settings**

| | |
|---|---|
| Population model | $(\mu_o + \mu_c, \lambda_o + \lambda_c)$ |
| Mutation type | standard, $n_\sigma = N$, $\tau_0 = (\sqrt{2n})^{-1}$ and $\tau_1 = (\sqrt{2\sqrt{n}})^{-1}$ |
| Crossover type | intermediate, $\rho = 2$ |
| Selection type | comma, $\kappa$, two-step-selection sex o |
| Initialization | $\sigma_i = \frac{|x^{(0)} - x^*|}{N}$ |
| Termination | fitness stagnation, $\theta = 10^{-12}$ |
| Constraint handling | TSES |
| Runs | 25 |

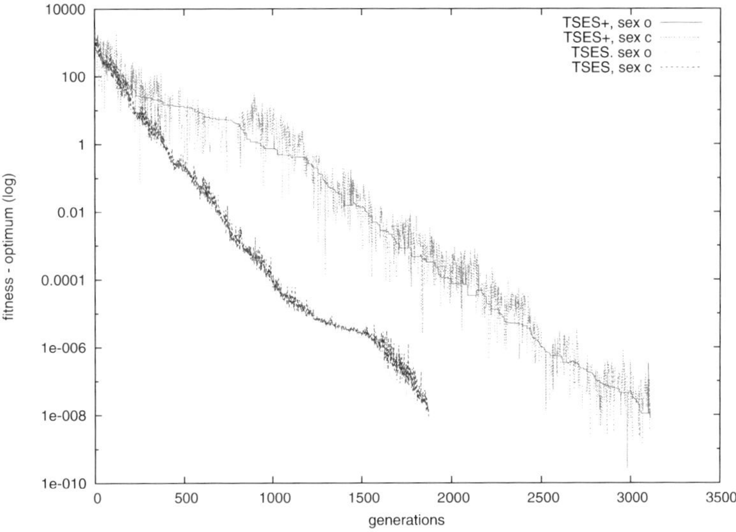

**Fig. 7.7.** Fitness development of typical runs of the TSES and the modification TSES+ on problem 2.40. The figure shows the fitness development of both sexes on a logarithmic scale. The TSES+ is faster and its sex $o$ is less fluctuating.

On problem g12 we observe a high quality of results with regard to the accuracy of the data structure. The (8+8,15+85)-, the (8+8,13+87)- as well as the (8+8,10+90)-TSES find the optimum with satisfying accuracy. The only exception is the (8+8, 13+87)-TSES with a fixed initial starting point. This result can be explained with outliers, as the neighboring sex ratios are successful. Again, we observe a saving of constraint function calls in comparison to death penalty.

Figure 7.7 shows a typical run of the TSES on problem 2.40 on a logarithmic scale. The negative fitness of sex $o$ is shown, since the individuals are located on the infeasible side of the optimum. The $o$-individuals are oscillating during the optimization, because they are *jumping* onto the infeasible side in every step. The figure also shows a TSES variant with a two step selection operator for sex $o$, the TSES+. It shows significant performance wins. The selection operator is modified as follows: at first, $\mu_o^\alpha$ individuals with sex $o$ are selected by objective function. In the second step, $\mu_o^\beta < \mu_o^\alpha$ individuals of the $\mu_o^\alpha$ ones are selected by fulfillment of constraints. The fastest run of the TSES+ reached the optimum within 2117 generations, the TSES needed 3830 generations. The median could be improved from 3219 to 5320 generations. The variance among the TSES+ $o$-individuals in each generation is smaller than the variance among the TSES $o$-individuals. Obviously, the $o$-individuals jump only slightly across the constraint boundary and thus enable a faster approximation of the optimum.

## The TSES on All Functions

Table 7.5 shows the experimental results of the TSES on all considered problems. While DP completely fails on Schwefel's constrained problem 2.40 the $(8+8,13+87)$-TSES reaches the optimum in every run. The same behavior can be observed on Schwefel's problem 2.41. In contrast to these problems the TSES is not able to approximate the optimum of problem TR2 arbitrarily. But at least a significant improvement in comparison to the results with DP can be observed. The behavior of the TSES on g01 has already been described. Despite an increase of the population sizes, the TSES could not achieve sufficient results on problem g02. On problem g04 a $(8+8,15\text{-}85)$-TSES failed, but the TSES with an offspring sex ratio of $13+87$ and $10+90$ found the optimum in every run. Again, the improvement in comparison to death penalty is significant. On problem g06, a highly constrained problem with a ratio of only 0.0066% of feasible search space, all algorithms, including death penalty were successful. Similar to problem g02 an increase of population sizes could not obtain promising approximation of the optimum of problem g07. The TSES could improve the results on problem g08 in comparison to death penalty, but demands a higher number of *ffc* and *cfc*. The results of the TSES on g09 were not satisfying, but slightly better than DP. Again, this improvement has to be paid with approximately 5 to 10 times higher number of *ffc* and *cfc*. On problem g11 various experiments around the sex ratio $(8+8,13+87)$ like recommended by Kramer and Schwefel [75] could not achieve promising results. Tests around the ratio $(8+8,10+200)$ were more successful. All experiments with the $(8+8,10+200)$-TSES and $\kappa = 200$ or $\kappa = 300$ showed sufficient results. A higher $\kappa$ causes a loss of efficiency. We have to emphasize that DP shows comparable results. But the number of constraint function calls *cfc* is about ten times higher than the *cfc* of the TSES. So, we can observe an

**Table 7.5.** Comparison of experimental results of the TSES on our constrained test problems. In most cases the TSES shows satisfying approximation results.

| | TSES | $\kappa$ | best | avg | worst | dev |
|---|---|---|---|---|---|---|
| 2.40 | (8+8,13+87) | 50 | −5000.0000000000 | −4999.9999999999 | −4999.9999999997 | 3.3E-7 |
| 2.41 | (8+8,13+87) | 50 | −17857.1428571482 | −17857.1428571426 | −17857.1428571374 | 8.5E-10 |
| TR2 | (8+8,13+87) | 300 | 2.0000000000 | 2.0000000095 | 2.0000000633 | 1.1E-8 |
| g01 | (40+40,50+400) | 50 | −14.9999999999 | −14.9999999999 | −14.9999999999 | 5.3E-15 |
| g02 | (40+40,50+400) | 50 | −0.7926079090 | −0.6583469345 | −0.4352731883 | 9.7E-3 |
| g04 | (8+8,13+87) | 50 | −30665.5386717833 | −30665.5386717833 | −30665.5386717832 | 1.7E-11 |
| g06 | (8+8,10+90) | 50 | −6961.8138755801 | −6961.8138755801 | −6961.8138755801 | 8.6E-13 |
| g07 | (40+40,50+400) | 50 | 24.3171014481 | 24.4613633236 | 24.8935876076 | 1.0E-2 |
| g08 | (8+8,15+85) | 50 | −0.0958250414 | −0.0958250414 | −0.0958250414 | 5.6E-18 |
| g09 | (8+8,13+87) | 50 | 680.6303038768 | 680.6345393750 | 680.6530448662 | 4.6E-4 |
| g11 | (8+8,10+200) | 300 | 0.7499900000 | 0.7499900169 | 0.7499903953 | 4.4E-9 |
| g12 | (8+8,10+90) | 50 | −1.0000000000 | −0.9999999999 | −0.9999999999 | 7.4E-16 |
| g16 | (40+40,50+400) | 50 | −1.9051552585 | −1.9051552585 | −1.9051552585 | 1.6E-15 |
| g24 | (8+8,10+90) | 50 | −5.5080132715 | −5.5080132715 | −5.5080132715 | 4.3E-13 |

efficiency advantage of the TSES. The experiments on problem g12 have already been commented in this section. On problem g16 the (8+8,13+87)-TSES is not more successful, but less efficient than DP with an average accuracy of 3 decimal places. But the (40+40,50+400)-TSES reaches the optimum with arbitrary accuracy demanding 3-4 times more *ffc* and *cfc*. The (8+8,10+90)-TSES is the best compromise between quality of the results and efficiency on problem g24. But we recommend to use DP as it demands much less *ffc* and *cfc*.

## 7.6  The Nested Angle Evolution Strategy

The last proposed constraint handling heuristic is the NAES. It is based on a nested ES and rotates the mutation ellipsoid until the success rates near the constraint boundary improve.

### 7.6.1  Meta-evolution for Mutation Ellipsoid Rotation

The success rates at the boundary of the feasible search space will change considerably, if the mutation ellipsoid is rotated by the angle $\delta$, see Figure 7.8. Here, $\delta$ is the smaller angle between one parameter axis and the constraint boundary. The mutation ellipsoid can adapt to a situation, where the infeasible search space does not cut off the success area. The success rate *improves* to $p_s = 0.5$ and consequently prevents premature step size reduction. This consideration leads to NAES. Figure 7.8 shows the situation when the mutation ellipsoid is rotated by angle $-\delta$. Rotation of the mutation ellipsoid can be achieved by correlated mutations introduced by Schwefel [132]. But experiments with the correlated mutations show that the diversity in the population of a $(15, 100)$-ES may be too low to achieve the adaptation of both mean step sizes and angles, see table 7.6.

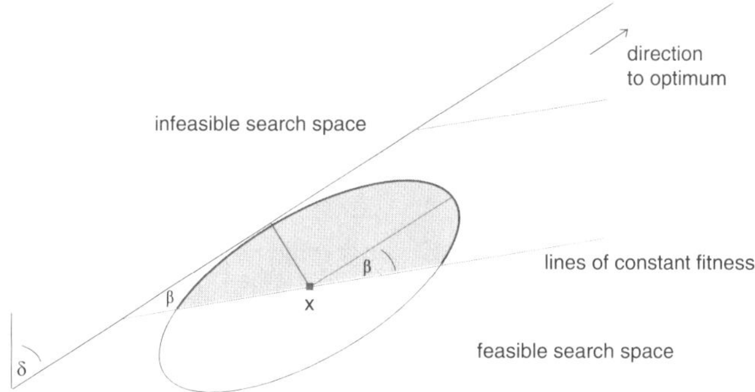


**Fig. 7.8.** Situation at the boundary of the feasible search space after rotation of the mutation ellipsoid by the angle $-\delta$: The self-adaptation process enables the step sizes $\sigma_1$ and $\sigma_2$ to form an ellipsoid that is not cut off by the constrained search space.

**Table 7.6.** The (15,100)-ES with correlated mutations and standard parameter settings on problem 2.40. The algorithm is far away from approximating the optimum in every run.

|      | best | mean | worst | dev | ffc | cfc |
|------|------|------|-------|-----|-----|-----|
| 2.40 | -5000.00000000 | -4942.9735056 | -4704.2122103 | 88.653 | 78 292 | 180 450 |

In our new nested angle evolution strategy (NAES) the outer ES adapts the angles for the rotation of the mutation ellipsoid of the inner ES. Here we use the advanced notation for nested ES introduced by Rechenberg [115]. In the $[\mu'/\rho' \,\overset{+}{,}\, \lambda'(\mu/\rho \,\overset{+}{,}\, \lambda)^\gamma]$-ES $\lambda'$ inner ES run for $\gamma$ generations, also called isolation time. In our approach the inner ES terminates based on fitness stagnation, i.e. when premature step size reduction occurs.

### 7.6.2 Experimental Analysis

Table 7.7 shows the experimental results of the $[5/2, 50(5/2, 50)]$-NAES on problems 2.40, 2.41, TR2 and g04 after 15 runs. The variables of the outer ES, which are the angles for the inner ES, are initialized within the interval $[0, \pi/2]$. The corresponding initial step sizes are $\sigma_i = \pi/8$ for all $i$. Mutation and recombination parameters are chosen as usual. For both ES fitness stagnation is chosen as termination condition. For all inner ES the setting $\theta = 10^{-9}$ and for the outer $\theta = 10^{-7}$ is chosen.

As the worst fitness values and the standard deviations show, the NAES is able to find the optimum of all problems with the accuracy the termination condition allows. The NAES is able to adapt the rotation angles and thus increases the success probability $p_s$. This even holds for problem TR2, whose angle $\delta$ and $p_s$ respectively, decrease rapidly during the convergence to the optimum. The NAES causes a very high number of constraint and fitness function calls as we expect from the nature of a meta-evolutionary approach. Every solution of the outer ES requires a full run of a couple of inner ES. But the NAES is the only constraint-handling method which is able to approximate the optimum of every type of constrained problem. In practice, the NAES might be too inefficient, but the performance can be improved by adequate parameter settings for the

**Table 7.7.** The $[5/2, 50(5/2, 50)]$-NAES on problems TR2, 2.40, 2.41 and g04. On all problems and in all runs the NAES is able to find the optimum with the desired accuracy.

|      | best | mean | worst | dev | ffc | cfc |
|------|------|------|-------|-----|-----|-----|
| 2.40 | -5000.000 | -5000.000 | -5000.000 | 1.2E-8 | 35 935 916 | 149 829 046 |
| 2.41 | -17857.143 | -17857.143 | -17857.143 | 3.8E-8 | 12 821 710 | 41 720 881 |
| TR2 | 2.000 | 2.000 | 2.000 | 3.1E-16 | 927 372 | 1 394 023 |
| g04 | -31025.560 | -31025.560 | -31025.560 | 2.1E-8 | 19 730 306 | 60 667 984 |

parameters $\mu, \mu', \lambda$ and $\lambda'$: further experiments showed that we can reduce the number of fitness function calls by decreasing the population sizes, e.g. with a $[3/2, 15(3/2, 15)]$-NAES. But smaller population sizes are not always sufficient to guarantee the diversity in the outer population that is necessary to find the correct rotation angles.

### 7.6.3  Outlook: Covariance Matrix Adaptation by Feasibility Classification

In the previous section 7.6 we introduced the idea of rotating the axes of the mutation ellipsoid parallel to the constraint boundary. But the angle of the rotation can also be calculated: similar to the CMA-ES [54] we propose to sample the search space at the constraint boundary and calculate the hyperplane that separates feasible from infeasible points. On way to learn the separating hyperplane is the perceptron algorithm [120]. Transforming the feasible and infeasible sample points $\mathbf{x}_j$ into the sample points $\mathbf{z}_j = (x_1, \ldots, x_N, 1)$ the perceptron algorithm is looking for the N+1 dimensional vector $\mathbf{w}$ that separates the two classes $\mathbf{z}_i \mathbf{w}^T > 0$ for the feasible points from and $\mathbf{z}_i \mathbf{w}^T < 0$ for the infeasible ones. The algorithm terminates after correct classification or after $k_{\max}$ iterations. The output $\mathbf{w}$ is an approximation of a hyperplane, which lies tangential to the infeasible region. In the next step we propose to adapt the mutation ellipsoid such that its principal axes lie parallel to the hyperplane. If all samples are feasible, the ES or CMA-ES works in its basic form. But if the number of infeasible samples is bigger than one, we rotate the mutation ellipsoid until it is parallel to the separating hyperplane. At least in the case of linear constraints we think the proposed sketch of a constrained handling technique might me successful.

## 7.7  Summary

1. Constraint handling for EC offers a huge potential for heuristics. The premature step size reduction and the resulting premature fitness stagnation could be shown experimentally for two traditional constraint-handling techniques, death penalty and a dynamic penalty function.
2. We proved step size reduction for a $(1+1)$-EA under simplified conditions. The proof is based on a success rate analysis considering a simplified EA model on linear functions. The situation at the constraint boundary can be modeled by two different states. From the success rates and the possible state transitions, the expected step size changes in every step can be derived.
3. Three new constraint-handling techniques were introduced with the aim of preventing an ES to suffer from premature fitness stagnation. We analyzed the two heuristics TSES and DSES experimentally on known constrained problems from literature and compared effectivity and efficiency.
4. The DSES is based on a least mutation strength $\epsilon$ together with an adaptation technique to reduce $\epsilon$. The parameters $\varpi$ and $\vartheta$ define the speed of the $\epsilon$-reduction process. Before applying the DSES it has to be considered,

whether it is worth to invest the higher number of fitness and constraint function calls.

5. The TSES is inspired by the concepts of sex and pairing. Each individual is assigned with a feature called sex which determines its selection objective. Modifications are necessary to prevent a step size explosion. The experiments revealed that the TSES is able to approximate the optimum in most of the cases. Sometimes a performance win in comparison to DP or the DSES was observed. An advantage of the TSES is that no infeasible starting points have to be available at the beginning of the search. A modification, the TSES+ was introduced, which applies the two step selection operator to the sex $o$. The TSES+ shows considerable performance improvements.

6. The proposed methods depend on new parameters, which have to be adapted carefully. But this drawback can be weakened. Population sizes have to be defined for almost every EA. The success of the TSES depends on the sex ratios, but here we offer examples for successful population ratios. The DSES only depends on two new parameters, $\varpi$ and $\vartheta$, which represent the speed of the $\epsilon$-reduction.

7. The NAES evolves the mutation ellipsoid rotation with a nested EA and consequently shows a poor performance. Hence, we cannot recommend the NAES for practical constraint handling. But the approach shows that the rotation of the mutation ellipsoid improves the success rate and may be a source for prospective constraint handling heuristics.

# Part IV

Summary

# 8  Summary and Conclusion

## 8.1  Contributions of This Book

Evolution created manifold creatures since its beginning with the first organisms, the prokaryotes, three to four billion years ago. Nowadays, computer scientists translate the principles of inheritance, variation and natural selection into algorithmic concepts known as EC. They mainly perform randomized search in the domain space. Among its most famous variants are GAs, ES, EP and GP growing more and more together. In the past many heuristic extensions have been proposed for EAs. The contribution of research in EC is comparable to old-fashioned AI research: the reduction of exponentially large search spaces by appropriate representations, operators and heuristics. The success of the search is often a result of the knowledge the practitioner integrates into the heuristics, a symptom of the no free lunch theorem. Sometimes the practitioner should be aware that classic search techniques may be faster. But most classical techniques demand domain knowledge to be applicable or require mathematical features like steadiness or the existence of the first or second derivative, which is not the case for many practical problems. Nevertheless, randomized search has grown to a strong, practicable and well-understood search technique, which is used all over the world in research and engineering. EAs have successfully been applied to domains like control, pattern recognition or automatic design. In the following, we summarize the main results and research contributions of this book.

- **Taxonomy of parameter adaptation.** The success of EAs partly depends on proper parameter settings. Various kinds of parameter adaptation techniques exist. We proposed an extended taxonomy for parameter setting, which is based on the early taxonomy of Eiben [37]. At first, the methods can be classified into *tuning before* and *controlling during* the optimization run. The latter can furthermore be classified into deterministic, adaptive, self-adaptive and meta-evolutionary approaches. Deterministic parameter control means changing the parameters in terms of the number of generations. Heuristic rules are the basis for adaptive parameter control.

O. Kramer: Self-Adaptive Heuristics for Evolutionary Computation, SCI 147, pp. 143–146, 2008.
  

- **Estimation of distribution view on self-adaptation.** Self-adaptation is an implicit parameter adaptation technique enabling the evolutionary search to tune the strategy parameters automatically by evolution. From the point of view of the EDA approach, we define self-adaptation as the evolution of optimal mixture distribution properties. This is very obvious for the mutation strength control of ES. They consist of a population of Gaussian distributions that estimate the optimum's location. A current population represents the present and therefore potentially *best* solutions. Self-adaptation of step sizes lets an ES converge to the optimum of many continuous optimization problems with arbitrary accuracy. In this work we introduced self-adaptive strategies that influence other distribution parameters.

For EC various genetic operators exist. ES exhibit plenty of mutation operators, from isotropic Gaussian mutation to the derandomized CMA. Most self-adaptive operators concern step size control of ES and EP, i.e. the variance of the mutation distribution function.

- **Introduction of new biased mutation variants.** Unlike the principle of unbiasedness of mutations, the BMO was introduced as a self-adaptive operator for biasing mutations into beneficial directions. The operator makes use of a self-adaptive bias coefficient vector $\xi$, which determines the direction of the bias. The absolute bias is computed by multiplication with the step sizes. A number of variants like the sBMO and cBMO have been introduced. Another variant is the DMO, whose bias is not controlled self-adaptively, but with the help of the descent direction between two successive populations' centers. It is based on the assumption of locality, i.e. the estimated direction to the global optimum can be derived from this local information. The obvious intuition could be shown by theoretical considerations, i.e. if the bias points into the descent direction, biased mutation is superior to unbiased mutation on monotone functions. Experimental analysis and statistical tests on typical test problems showed the superiority of biased mutation on multimodal functions like rosenbrock, rosenbrock with noise and on ridge functions like parabolic ridge and sharp ridge. The mutation of the bias coefficient $\xi$ with settings around $\gamma \approx 0.1$ revealed the best experimental results. The same holds for population ratios around (15,100).
- **Application of biased mutation to constrained search spaces.** Our experimental analysis revealed that biased mutation is appropriate for searching at the boundary of constrained search spaces. The experiments on constrained functions like Schwefel's problem 2.40 or g04 from the g-library of constrained search spaces showed that the BMO in combination with death penalty exhibits better convergence properties than the ES Gaussian standard mutation. The latter suffers from premature stagnation at the constraint boundary. Statistical tests confirmed the significance of the results.

GAs in the classical form seldom use self-adaptation. On the one hand this is the result of the historical development, on the other hand classical GAs work with discrete representations. For these not many successful self-adaptation techniques are known.

- **Self-adaptive inversion mutation.** In this work we presented the SA-INV, a successful example for discrete mutation strength self-adaptation. Inversion mutation is the most common mutation operator for GAs on combinatorial problems like the TSP. SA-INV is the self-adaptive variant of inversion mutation. Its strategy parameter $k$ determines the number of successive inversions, i.e. swaps of edges. A convergence analysis proves that INV and SA-INV find the optimum in finite time with probability one. The experimental analysis showed that SA-INV is able to speed up the optimization, in particular at the beginning of the search. Parameter $k$ is usually decreasing during the optimization as less and less edge swaps are advantageous for the improvement and not the destruction of solutions. But when reaching $k \approx 1$, SA-INV slows down, as self-adaptation still produces solutions with $k > 1$ and a worse success rate. We call this problem *strategy bound problem*. To overcome the *strategy bound problem* in later search phases, we introduced the SA-INV-c heuristic, which sets $k = 1$ constantly if the condition $u \geq \frac{\lambda}{2}$ is fulfilled. The experimental analysis shows that the modification improves the performance of SA-INV. The Wilcoxon rank-sum test validates the statistical significance of the results.

Two successful examples for self-adaptive mutations have been introduced. But also self-adaptive properties for the second important genetic operator, crossover, have been investigated. Crossover plays a controversial discussed role within EAs. BBH and GR hypothesis give conflictive explanations for the working principle of crossover. The structure of most existing crossover operators is not adaptive, but either random or fixed.

- **Self-adaptation of structural crossover properties.** We introduced self-adaptive n-point crossover, self-adaptive partially mapped crossover and self-adaptive recombination for ES. Our experiments showed that self-adaptive crossover achieves no significant improvements on combinatorial problems like the TSP (PMX), on continuous optimization problems (discrete and intermediate recombination) as well as on bit-string problems (one- and n-point crossover). We see two main reasons for the *weakness* of self-adaptive crossover. First, the fitness gain achieved by *optimal* crossover settings is weak, e.g. for multimodal continuous problems. Experiments with the crossover optimization EA could only discover valuable fitness gain on unimodal or monotone functions like sphere and ridge. Second, even if the gain of a specific crossover strategy parameter set is advantageous, it may not be optimal in the following generation. This hinders a self-adaptation of useful crossover points and *building blocks*. For future research we see the challenge to tighten the link between strategy parameter adaptation and fitness gain by choosing individuals from different mating pools or subpopulations with different properties.

Self-adaptation controls the estimated distribution of the optimum's location. But it also suffers from shortcomings. One example for the latter is the premature step size reduction of ES in the vicinity of infeasible parts of the search space. When self-adaptation fails, *adaptive* parameter control heuristics can help as the proposed constraint handling methods of this book show.

- **Constraint handling heuristics for optima with active constraints.** We proved step size reduction for a (1+1)-EA leading to premature convergence under simplified conditions, i.e. linear fitness function and constraints. Our proof is based on a success rate analysis considering a simplified EA model. The situation at the constraint boundary can be modeled by two different states. From the success rates and the possible state transitions, the expected step size changes can be derived for each step. Experiments showed the premature fitness stagnation for two traditional constraint-handling techniques, death penalty and a dynamic penalty function on constrained problems. In this work we introduced three constraint-handling techniques aiming at the premature fitness stagnation of ES. The DSES makes use of a least mutation strength $\epsilon$. But it also prevents the convergence of the optimum. Hence, we introduced an adaptation technique to reduce $\epsilon$ and allow the unlimited convergence. The parameters $\varpi$ and $\vartheta$ define the speed of the $\epsilon$-reduction process. The second heuristic is the TSES, inspired by the concepts of sex and pairing. Each individual is assigned with a feature called sex, which determines its selection objective. Some modifications are necessary to prevent a step size explosion. Last, the NAES evolves the covariance matrix with a nested EA and consequently shows a poor performance. Hence, we cannot recommend the NAES for practical constraint handling, but the approach shows that the rotation of the mutation ellipsoid improves the success rate. Extensive experimental analysis showed the behavior and properties of the proposed methods.

## 8.2  Conclusion

Self-adaptation is an efficient way of improving evolutionary search. But a necessary condition for its success is a tight link between strategy parameters and fitness. This link is not existing for every type of parameter. The crossover points are an example for a weak link. Mutation strongly controls the protagonist exploration and antagonist exploitation and therefore offers a direct link between its strategy parameters and the corresponding fitness. The bias control of the BMO and the number of successive edge swaps of SA-INV are examples for the success of self-adaptation of mutation parameters. But self-adaptation can also be misleading: due to local optima in the strategy parameter space, it may result in premature fitness stagnation. Hence, it has to be used with care and sense for search domain characteristics. Furthermore, adaptive heuristic extensions like the proposed constraint handling techniques offer useful means to overcome premature stagnation. To summarize all results, well-designed self-adaptive and adaptive heuristics are efficient techniques to improve evolutionary search and convergence properties.

# Part V

# Appendix

# A  Continuous Benchmark Functions

## A.1  Unimodal Numerical Functions

**Sphere Model**

Minimize

$$f(\mathbf{x}) = \sum_{i=1}^{N} x_i^2 \quad \text{with } \mathbf{x} \in \mathbb{R}^N \tag{A.1}$$

with properties:

- unimodal, separable
- scalable
- $\mathbf{x} \in [-100, 100]^N$,

minimum

$$\mathbf{x}^* = (0, \ldots, 0)^T, \text{ with } \quad f(\mathbf{x}^*) = 0.$$

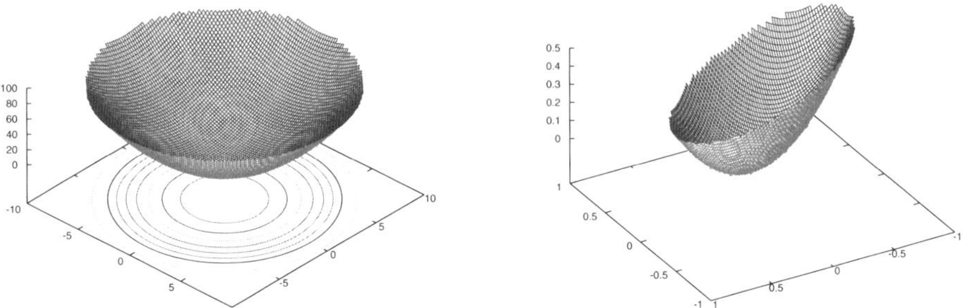

**Fig. A.1.** Left: plot of the sphere function. Right: Plot of the double sum function (Schwefel's problem 1.2, [135])

O. Kramer: Self-Adaptive Heuristics for Evolutionary Computation, SCI 147, pp. 149–158, 2008.
springerlink.com                                    © Springer-Verlag Berlin Heidelberg 2008

**Double Sum**

Minimize

$$f(\mathbf{x}) = \sum_{i=1}^{N} \left( \sum_{j=1}^{i} (x_j) \right)^2 \qquad \text{with } \mathbf{x} \in \mathbb{R}^N \qquad (A.2)$$

with properties

- unimodal, non-separable
- scalable
- $\mathbf{x} \in [-100, 100]^N$,

minimum $\mathbf{x}^* = (0, \ldots, 0)^T$ with $f(\mathbf{x}^*) = 0$.

## A.2  Multimodal Numerical Functions

**Rosenbrock**

Minimize

$$f(\mathbf{x}) = \sum_{i=1}^{n-1} \left( (100(x_i^2 - x_{i+1})^2 + (x_i - 1)^2) \right) \qquad \text{with } \mathbf{x} \in \mathbb{R}^N \qquad (A.3)$$

with properties:

- multi-modal for $N > 4$,
- non-separable, scalable
- very narrow valley from local optimum to global optimum
- $\mathbf{x} \in [-100, 100]^N$,

minimum $\mathbf{x}^* = (1, \ldots, 1)^T$ with $f(\mathbf{x}^*) = 0$. For higher dimensions the optimum exhibits a local optimum at $x = (-1, \ldots, 1)^T$.

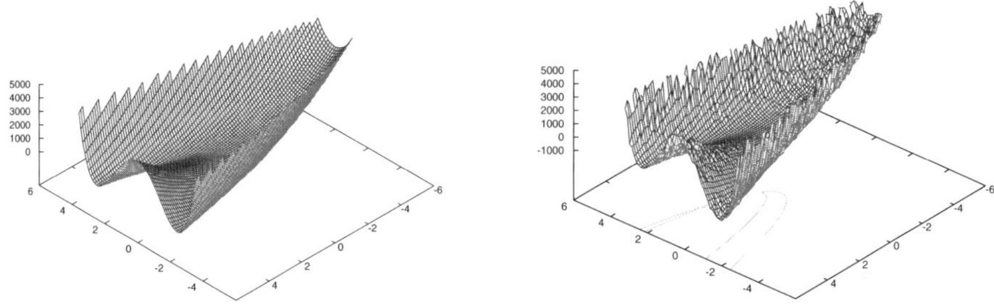

**Fig. A.2.** Plot of the rosenbrock function (left) and rosenbrock with noise in fitness (right)

**Rosenbrock with Noise in Fitness**

Minimize

$$f(\mathbf{x}) = \left( \sum_{i=1}^{n-1} \left( 100(x_i^2 - x_{i+1})^2 + (x_i - 1)^2 \right) \right) \cdot (1 + 0.4|\mathcal{N}(0,1)|) \quad \text{with } \mathbf{x} \in \mathbb{R}^N \tag{A.4}$$

minimum $\mathbf{x}^* = (1, \ldots, 1)^T$ with $f(\mathbf{x}^*) = 0$. For higher dimensions the optimum exhibits a local optimum at $x = (-1, \ldots, 1)^T$, with properties like rosenbrock.

**Rastrigin**

Minimize

$$f(\mathbf{x}) = \sum_{i=1}^{N} \left( x_i^2 - 10\cos(2\pi x_i) + 10 \right) \quad \text{with } \mathbf{x} \in \mathbb{R}^N \tag{A.5}$$

with properties
- multi-modal, separable
- huge number of local optima
- scalable
- $\mathbf{x} \in [-5,5]^N$,

minimum $\mathbf{x}^* = (0, \ldots, 0)^T$, with $\quad f(\mathbf{x}^*) = 0$.

**Griewank**

Minimize

$$f(\mathbf{x}) = \sum_{i=1}^{N} \frac{x_i^2}{4000} - \prod_{i=1}^{N} \cos\left( \frac{x_i}{\sqrt{i}} \right) + 1 \quad \text{with } \mathbf{x} \in \mathbb{R}^N \tag{A.6}$$

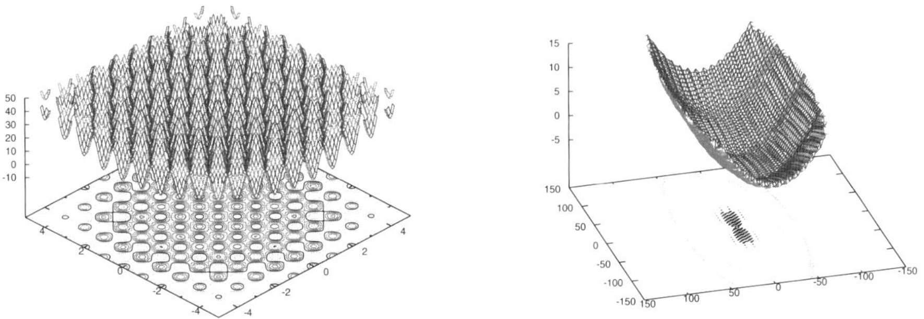

**Fig. A.3.** Left: plot of the rastrigin function. Right: plot of the griewank function.

with properties
- multi-modal, non-separable
- scalable,

minimum $\mathbf{x}^* = (0,\dots,0)^T$, with   $f(\mathbf{x}^*) = 0$.

## A.3  Ridge Functions

### Sharp Ridge

Minimize

$$f(\mathbf{x}) = -x_1 + 100\sqrt{\sum_{i=2}^{N} x_i^2} \quad \text{with } \mathbf{x} \in \mathbb{R}^N \tag{A.7}$$

One coordinate is linearly added to the function. The function is converging towards infinity. $\mathbf{x}^0 \in [-100, 100]^N$.

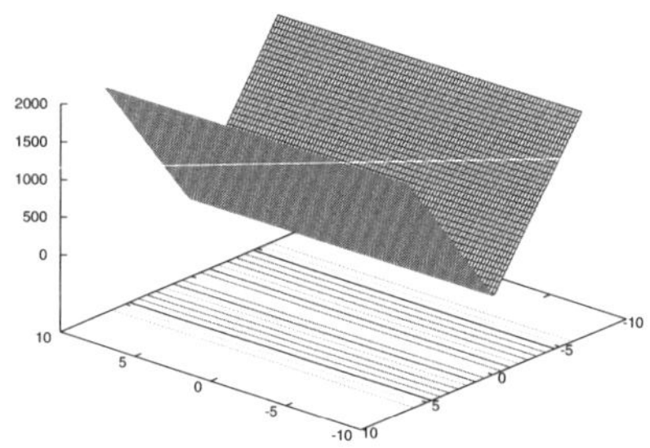

**Fig. A.4.** Plot of the sharp ridge function

### Parabolic Ridge

Minimize

$$f(\mathbf{x}) = -x_1 + \sum_{i=2}^{N} x_i^2 \quad \text{with } \mathbf{x} \in \mathbb{R}^N \tag{A.8}$$

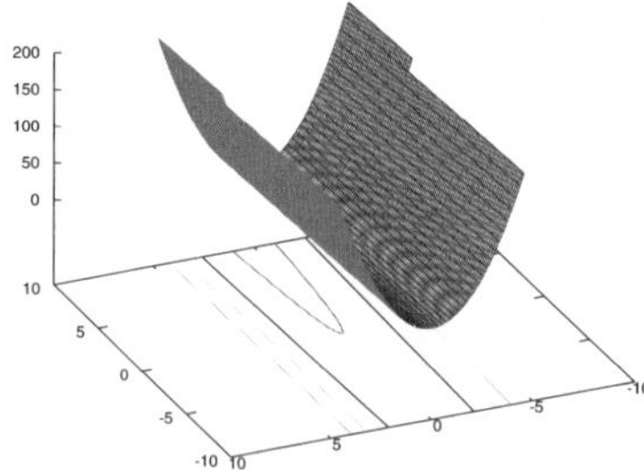

**Fig. A.5.** Plot of the parabolic ridge function

One coordinate is linearly added to the function. The function is converging towards infinity. $\mathbf{x}^0 \in [-100, 100]^N$.

## A.4  Numerical Functions in Bit String Representations

**Onemax**

Onemax is a binary function. For a bit string $\mathbf{x} \in \{0, 1\}^N$, maximize

$$f(\mathbf{x}) = \sum_{i=1}^{N} x_i \quad \text{with } \mathbf{x} \in \{0, 1\}^N \tag{A.9}$$

optimum $\mathbf{x}^* = (1, \ldots, 1)^T$ with $f(\mathbf{x}^*) = N$.

**Ackley**

Minimize

$$f(\mathbf{x}) = -20 \cdot \exp\left(-0.2\sqrt{\frac{1}{N}\sum_{i=1}^{N} x_i^2}\right) - \exp\left(\frac{1}{N}\sum_{i=1}^{N}\cos(2\pi x_i)\right) + 20 + e \tag{A.10}$$

with properties

- multi-modal
- huge number of local optima

and optimum $\mathbf{x}^* = (0, \ldots, 0)^T$ with $f(\mathbf{x}^*) = 0$.

## A.5  Constrained Numerical Functions

**TR - Tangent Problem**

Minimize

$$f(\mathbf{x}) = \sum_{i=1}^{N} x_i^2 \qquad \text{(n-dim. sphere model)} \tag{A.11}$$

constraints

$$g(\mathbf{x}) = \sum_{i=1}^{N} x_i - t > 0, \qquad t \in \mathbb{R} \qquad \text{(tangent)} \tag{A.12}$$

For n=k and t=k the minimum lies at:

$$\mathbf{x}^* = (1, \ldots, 1)^T, \quad \text{with} \quad f(\mathbf{x}^*) = k. \tag{A.13}$$

TR2 is TR with $N = 2$ and $t = 2$.

**Schwefel's Problem 2.40**

Minimize

$$f(\mathbf{x}) = -\sum_{i=1}^{5} x_i \tag{A.14}$$

constraints

$$g_j(\mathbf{x}) = \begin{cases} x_j \geq 0, & \text{for } j = 1, \ldots, 5 \\ -\sum_{i=1}^{5}(9+i)x_i + 50000 \geq 0, & \text{for } j = 6 \end{cases} \tag{A.15}$$

minimum $\mathbf{x}^* = (5000, 0, 0, 0, 0)^T$ with $f(\mathbf{x}^*) = -5000$, $g_2$ to $g_6$ active.

**Schwefel's Problem 2.41**

Minimize

$$f(\mathbf{x}) = -\sum_{i=1}^{5}(ix_i) \tag{A.16}$$

constraints like problem 2.40.
minimum $\mathbf{x}^* = \left(0, 0, 0, 0, \frac{50000}{14}\right)^T$ with $f(\mathbf{x}^*) = -\frac{250000}{14}$,
$g_j$ active for $j = 1, 2, 3, 4, 6$.

**g01**

Minimize

$$f(\mathbf{x}) = 5\sum_{i=1}^{4}(x_i) - 5\sum_{i=1}^{4}(x_i^2) - \sum_{i=5}^{13}(x_i) \tag{A.17}$$

constraints:

$$\begin{aligned}
g_1(\mathbf{x}) &= 2x_1 + 2x_2 + x_{10} + x_{11} - 10 \leq 0 \\
g_2(\mathbf{x}) &= 2x_1 + 2x_3 + x_{10} + x_{12} - 10 \leq 0 \\
g_3(\mathbf{x}) &= 2x_2 + 2x_3 + x_{11} + x_{12} - 10 \leq 0 \\
g_4(\mathbf{x}) &= -8x_1 + x_{10} \leq 0 \\
g_5(\mathbf{x}) &= -8x_2 + x_{11} \leq 0 \\
g_6(\mathbf{x}) &= -8x_3 + x_{12} \leq 0 \\
g_7(\mathbf{x}) &= -2x_4 - x_5 + x_{10} \leq 0 \\
g_8(\mathbf{x}) &= -2x_6 - x_7 + x_{11} \leq 0 \\
g_9(\mathbf{x}) &= -2x_8 - x_9 + x_{12} \leq 0
\end{aligned} \tag{A.18}$$

feasible intervals $0 \leq x_j \leq 1$ for $j = 1, \ldots, 9$, $\quad 0 \leq x_j \leq 100$ for $j = 10, 11, 12$
$0 \leq x_{13} \leq 1$, minimum $\mathbf{x}^* = (1, 1, 1, 1, 1, 1, 1, 1, 1, 3, 3, 3, 1)^T$ with $f(\mathbf{x}^*) = -15$,
feasible starting point $\mathbf{x}^{(0)} = (0.5, \ldots, 0.5)^T$ with $\quad f(\mathbf{x}^{(0)}) = 0.5$.

**g02**

Minimize

$$f(\mathbf{x}) = -\left| \frac{\sum_{i=1}^{N} \cos^4(x_i) - 2 \prod_{i=1}^{N} \cos^2(x_i)}{\sqrt{\sum_{i=1}^{N} i x_i^2}} \right| \tag{A.19}$$

constraints:

$$\begin{aligned} g_1(\mathbf{x}) &= 0.75 - \prod_{i=1}^{N} x_i \le 0 \\ g_2(\mathbf{x}) &= \sum_{i=1}^{N} x_i - 7.5N \le 0 \end{aligned} \tag{A.20}$$

with N=20, feasible intervals $0 < x_j \le 10$ for $j = 1, \ldots, N$, best known value $\mathbf{x}^* = (3.162, 3.128, 3.094, 3.061, 3.027, 2.993, 2.958, 2.921, 0.494, 0.488, 0.482, 0.476, 0.471, 0.466, 0.461, 0.456, 0.452, 0.448, 0.444, 0.440)^T$, $f(\mathbf{x}^*) = 0.80361910412559$, feasible starting point $\mathbf{x}^{(0)} = (1, \ldots, 1)^T$ with $f(\mathbf{x}^{(0)}) = 0.117616$.

**g04 - Himmelblau's Problem**

Minimize

$$f(\mathbf{x}) = 5.3578547 x_3^2 + 0.8356891 x_1 x_5 + 37.293239 x_1 - 40792.141 \tag{A.21}$$

constraints

$$\begin{aligned} g_1(\mathbf{x}) &= 85.334407 + 0.0056858 x_2 x_5 + 0.0006262 x_1 x_4 - 0.0022053 x_3 x_5 \\ g_2(\mathbf{x}) &= 80.51249 + 0.0071317 x_2 x_5 + 0.0029955 x_1 x_2 + 0.0021813 x_3^2 \\ g_3(\mathbf{x}) &= 9.300961 + 0.0047026 x_3 x_5 + 0.0012547 x_1 x_3 + 0.0019085 x_3 x_4 \\ 0 &\le g_1(\mathbf{x}) \le 92 \\ 90 &\le g_2(\mathbf{x}) \le 110 \\ 20 &\le g_3(\mathbf{x}) \le 25 \end{aligned} \tag{A.22}$$

with $78 \le x_1 \le 102$, $33 \le x_2 \le 45$ and $27 \le x_i \le 45$ ($i = 3, 4, 5$), minimum $\mathbf{x}^* = (78.000, 33.000, 29.995, 45.000, 36.776)^T$ with $f(\mathbf{x}^*) = -30665.53867178332$.

**g06**

Minimize

$$f(\mathbf{x}) = (x_1 - 10)^3 + (x_2 - 20)^3 \tag{A.23}$$

constraints

$$\begin{aligned} g_1(\mathbf{x}) &= -(x_1 - 5)^2 - (x_2 - 5)^2 + 100 \le 0 \\ g_2(\mathbf{x}) &= (x_1 - 6)^2 + (x_2 - 5)^2 - 82.81 \le 0 \end{aligned} \tag{A.24}$$

where $13 \le x_1 \le 100$ and $0 \le x_2 \le 100$, minimum $\mathbf{x}^* = (14.09500000000000064, 0.8429607892154795668)^T$ with $f(\mathbf{x}^*) = -6961.81387558015$, both constraints are active, feasible starting point $\mathbf{x}^{(0)} = (6, 9)^T$ with $f(\mathbf{x}^{(0)}) = -3250$.

## g07

Minimize

$$f(\mathbf{x}) = x_1^2 + x_2^2 + x_1 x_2 - 14x_1 - 16x_2 + (x_3 - 10)^2 + 4(x_4 - 5)^2 + (x_5 - 3)^2$$
$$+2(x_6 - 1)^2 + 5x_7^2 + 7(x_8 - 11)^2 + 2(x_9 - 10)^2 + (x_{10} - 7)^2 + 45$$
$$(\mathrm{A.25})$$

constraints

$$
\begin{aligned}
g_1(\mathbf{x}) &= -105 + 4x_1 + 5x_2 - 3x_7 + 9x_8 \le 0 \\
g_2(\mathbf{x}) &= 10x_1 - 8x_2 - 17x_7 + 2x_8 \le 0 \\
g_3(\mathbf{x}) &= -8x_1 + 2x_2 + 5x_9 - 2x_{10} \le 0 \\
g_4(\mathbf{x}) &= 3(x_1 - 2)^2 + 4(x_2 - 3)^2 + 2x_3^2 - 7x_4 - 120 \le 0 \\
g_5(\mathbf{x}) &= 5x_1^2 + 8x_2 + (x_3 - 6)^2 - 2x_4 - 40 \le 0 \\
g_6(\mathbf{x}) &= x_1^2 + 2(x_2 - 2)^2 - 2x_1 x_2 + 14x_5 - 6x_6 \le 0 \\
g_7(\mathbf{x}) &= 0.5(x_1 - 8)^2 + 2(x_2 - 4)^2 + 3x_5^2 - x_6 - 30 \le 0 \\
g_8(\mathbf{x}) &= -3x_1 + 6x_2 + 12(x_9 - 8)^2 - 7x_{10} \le 0
\end{aligned}
$$
$$(\mathrm{A.26})$$

where $-10 \le x_j \ge 10$ for $j = (1, \ldots, 10)$, optimum $\mathbf{x}^* = (2.171, 2.363, 8.773, 5.095, 0.990, 1.430, 1.321, 9.828, 8.280, 8.375)^T$ with $f(\mathbf{x}^*) = 24.30620906818$, feasible starting point $\mathbf{x}^{(0)} = (3, 1, 6, 10, 1, 4, 2, 1, 8, 8)^T$ with $f(\mathbf{x}^{(0)}) = 867$.

## g08

Minimize
$$f(\mathbf{x}) = -\frac{\sin^3(2\pi x_1) \sin(2\pi x_2)}{x_1^3(x_1 + x_2)} \tag{A.27}$$

constraints
$$
\begin{aligned}
g_1(\mathbf{x}) &= x_1^2 - x_2 + 1 \le 0 \\
g_2(\mathbf{x}) &= 1 - x_1 + (x_2 - 4)^2 \le 0
\end{aligned}
\tag{A.28}
$$

where $0 \le x_1 \le 10$ and $0 \le x_2 \le 10$, optimum $\mathbf{x}^* = (1.227, 4.245)^T$ with $f(\mathbf{x}^*) = -0.0958250414180359$, feasible starting point $\mathbf{x}^{(0)} = (1, 4)^T$ with $f(\mathbf{x}^{(0)}) = 2.879077639283126\mathrm{E}{-63}$.

## g09

Minimize

$$f(\mathbf{x}) = (x_1 - 10)^2 + 5(x_2 - 12)^2 + x_3^4 + 3(x_4 - 11)^2 + 10x_5^6 + 7x_6^2 + x_7^4 - 4x_6 x_7 - 10x_6 - 8x_7$$
$$(\mathrm{A.29})$$

constraints

$$
\begin{aligned}
g_1(\mathbf{x}) &= -127 + 2x_1^2 + 3x_2^4 + x_3 + 4x_4^2 + 5x_5 \le 0 \\
g_2(\mathbf{x}) &= -282 + 7x_1 + 3x_2 + 10x_3^2 + x_4^2 - x_5 \le 0 \\
g_3(\mathbf{x}) &= -196 + 23x_1 + x_2^2 + 6x_6^2 - 8x_7 \le 0 \\
g_4(\mathbf{x}) &= 4x_1^2 + x_2^2 - 3x_1 x_2 + 2x_3^2 + 5x_6 - 11x_7 \le 0
\end{aligned}
$$
$$(\mathrm{A.30})$$

with $-10 \leq x_j \leq 10$ for $j=(1,\ldots,7)$, optimum $\mathbf{x}^* = (2.330, 1.951, -0.477, 4.365,$ $-0.624, 1.038, 1.594)^T$ with $f(\mathbf{x}^*) = 680.630057374402$, feasible starting point $\mathbf{x}^{(0)} = (0,0,0,0,0,0,0)^T$ with $f(\mathbf{x}^{(0)}) = 1183.0$.

## g11

Minimize

$$f(\mathbf{x}) = -x_1^2 + (x_2 - 1)^2 \tag{A.31}$$

constraints

$$h(\mathbf{x}) = x_2 - x_1^2 = 0 \tag{A.32}$$

where $-1 \leq x_1 \leq 1$ and $-1 \leq x_2 \leq 1$, optimum $\mathbf{x}^* = (-0.707036070037170616,$ $0.00000004333606807)^T$ with $f(\mathbf{x}^*) = 0.7499$.

## g12

Minimize

$$f(\mathbf{x}) = -\left(100 - (x_1 - 5)^2 - (x_2 - 5)^2 - (x_3 - 5)^2\right)/100 \tag{A.33}$$

constraints

$$g(\mathbf{x}) = (x_1 - p)^2 + (x_2 - q)^2 - (x_3 - r)^2 - 0.0625 \leq 0 \tag{A.34}$$

"where $0 \leq x_i \leq 10$ $(i = 1, 2, 3)$ and $p, q, r = 1, 2, \ldots, 9$. The feasible region of the search space consists of $9^3$ disjoined spheres. A point $(x_1, x_2, x_3)^T$ is feasible, iff there exist $p, q, r$ such that the above inequality holds. The optimum is located at $\mathbf{x} = (5, 5, 5)^T$ where $f(\mathbf{x}) = -1$. The solution lies within the feasible region." [83]

## g16

Minimize

$$f(\mathbf{x}) = 0.000117 y_{14} + 0.1365 + 0.00002358 y_{13} + 0.000001502 y_{16} + 0.0321 y_{12}$$
$$+ 0.004324 y_5 + 0.0001 \frac{c_{15}}{c_{16}} + 37.48 \frac{y_2}{c_{12}} - 0.0000005843 y_{17}$$
$$\tag{A.35}$$

constraints

$$g_1(\mathbf{x}) = \frac{0.28}{0.72} y_5 - y_4 \leq 0$$
$$g_2(\mathbf{x}) = x_3 - 1.5 x_2 \leq 0$$
$$\vdots \tag{A.36}$$
$$g_{37}(\mathbf{x}) = 2802713 - y_{17} \leq 0$$
$$g_{38}(\mathbf{x}) = y_{17} - 12146108 \leq 0$$

where

$$y_1 = x_2 + x_3 + 41.6$$
$$c_1 = 0.024x_4 - 4.62$$
$$\vdots$$
$$c_{16} = 1.104 - 0.72y_{15}$$
$$c_{17} = y_9 + x_5$$

$$(A.37)$$

with bounds $704.4148 \leq x_1 \leq 906.3855$, $68.6 \leq x_2 \leq 288.88$, $0 \leq x_3 \leq 134.75$, $193 \leq x_4 \leq 287.0966$ and $25 \leq x_5 \leq 84.1988$, best known solution $\mathbf{x}^* \approx (705.174, 68.599, 102.899, 282.324, 37.584)^T$ where $f(\mathbf{x}^*) = -1.90515525853479$, constraints and conditions can be found in [83].

## g24

Minimize

$$f(\mathbf{x}) = -x_1 - x_2 \tag{A.38}$$

constraints

$$g_1(\mathbf{x}) = -2x_1^4 + 8x_1^3 - 8x_1^2 + x_2 - 2 \leq 0$$
$$g_2(\mathbf{x}) = -4x_1^4 + 32x_1^3 - 88x_1^2 + 96x_1 + x_2 - 36 \leq 0 \tag{A.39}$$

with $0 \leq x_1 \leq 3$ and $0 \leq x_2 \leq 4$, optimum $\mathbf{x}^* = (2.329520197, 3.178493074)^T$ with $f(\mathbf{x}^*) = -5.508013271$, feasible starting point $\mathbf{x}^{(0)} = (0,0)^T$ with $f(\mathbf{x}^{(0)}) = 0.0$.

## Properties of the g Problems

Table A.1 summarizes the properties of the g-problems as described in [83].

**Table A.1.** Properties of the g problems. N is the number of objective variables, $|F|/|S|$ is the estimated ratio between the feasible search space $F$ and whole search space $S$. LI is the number of linear inequality constraints, NI the number of nonlinear inequality constraints, LE is the number of linear equality constraints, NE is the number of nonlinear equality constraints, and *active* is the number of active constraints.

| problem | N | type of function | $|F|/|S|$ (%) | LI | NI | LE | NE | active |
|---|---|---|---|---|---|---|---|---|
| g01 | 13 | quadratic | 0.0111 | 9 | 0 | 0 | 0 | 6 |
| g02 | 20 | nonlinear | 99.9971 | 0 | 2 | 0 | 0 | 1 |
| g04 | 5 | quadratic | 52.1230 | 0 | 6 | 0 | 0 | 2 |
| g06 | 2 | cubic | 0.0066 | 0 | 2 | 0 | 0 | 2 |
| g07 | 10 | quadratic | 0.0003 | 3 | 5 | 0 | 0 | 6 |
| g08 | 2 | nonlinear | 0.8560 | 0 | 2 | 0 | 0 | 0 |
| g09 | 7 | polynomial | 0.5121 | 0 | 4 | 0 | 0 | 2 |
| g11 | 2 | quadratic | 0.0000 | 0 | 0 | 0 | 1 | 1 |
| g12 | 3 | quadratic | 4.7713 | 0 | 1 | 0 | 0 | 0 |
| g16 | 5 | nonlinear | 0.0204 | 4 | 34 | 0 | 0 | 4 |
| g24 | 2 | linear | 79.6556 | 0 | 2 | 0 | 0 | 2 |

# B  Discrete Benchmark Functions

## B.1  Traveling Salesman Problems

### Ulysses16

TSP problem of TSPlib from Reinelt [117]: coordinates of 16 locations of Odysseus' journey home to Ithaca, also known as Homer's Odyssey; geographical distances; length of optimal tour $f(\mathbf{x}^*) = 6859$.

**Table B.1.** Coordinates of problem *ulysses16*, geographical distances

| city | latitude | longitude | city | latitude | longitude |
|------|----------|-----------|------|----------|-----------|
| 1 | 38.24 | 20.42 | 9 | 41.23 | 9.10 |
| 2 | 39.57 | 26.15 | 10 | 41.17 | 13.05 |
| 3 | 40.56 | 25.32 | 11 | 36.08 | -5.21 |
| 4 | 36.26 | 23.12 | 12 | 38.47 | 15.13 |
| 5 | 33.48 | 10.54 | 13 | 38.15 | 15.35 |
| 6 | 37.56 | 12.19 | 14 | 37.51 | 15.17 |
| 7 | 38.42 | 13.11 | 15 | 35.49 | 14.32 |
| 8 | 37.52 | 20.44 | 16 | 39.36 | 19.56 |

### Berlin52

TSP problem of TSPlib from Reinelt [117]: coordinates of 52 locations in Berlin, Germany; Euclidean distances; length of optimal tour $f(\mathbf{x}^*) = 7542$.

O. Kramer: Self-Adaptive Heuristics for Evolutionary Computation, SCI 147, pp. 159–161, 2008.
springerlink.com                                      © Springer-Verlag Berlin Heidelberg 2008

**Table B.2.** Coordinates of problem *berlin52*, Euclidean distances

| city | x | y | city | x | y | city | x | y |
|---|---|---|---|---|---|---|---|---|
| 1 | 565.0 | 575.0 | 19 | 510.0 | 875.0 | 37 | 770.0 | 610.0 |
| 2 | 25.0 | 185.0 | 20 | 560.0 | 365.0 | 38 | 795.0 | 645.0 |
| 3 | 345.0 | 750.0 | 21 | 300.0 | 465.0 | 39 | 720.0 | 635.0 |
| 4 | 945.0 | 685.0 | 22 | 520.0 | 585.0 | 40 | 760.0 | 650.0 |
| 5 | 845.0 | 655.0 | 23 | 480.0 | 415.0 | 41 | 475.0 | 960.0 |
| 6 | 880.0 | 660.0 | 24 | 835.0 | 625.0 | 42 | 95.0 | 260.0 |
| 7 | 25.0 | 230.0 | 25 | 975.0 | 580.0 | 43 | 875.0 | 920.0 |
| 8 | 525.0 | 1000.0 | 26 | 1215.0 | 245.0 | 44 | 700.0 | 500.0 |
| 9 | 580.0 | 1175.0 | 27 | 1320.0 | 315.0 | 45 | 555.0 | 815.0 |
| 10 | 650.0 | 1130.0 | 28 | 1250.0 | 400.0 | 46 | 830.0 | 485.0 |
| 11 | 1605.0 | 620.0 | 29 | 660.0 | 180.0 | 47 | 1170.0 | 65.0 |
| 12 | 1220.0 | 580.0 | 30 | 410.0 | 250.0 | 48 | 830.0 | 610.0 |
| 13 | 1465.0 | 200.0 | 31 | 420.0 | 555.0 | 49 | 605.0 | 625.0 |
| 14 | 1530.0 | 5.0 | 32 | 575.0 | 665.0 | 50 | 595.0 | 360.0 |
| 15 | 845.0 | 680.0 | 33 | 1150.0 | 1160.0 | 51 | 1340.0 | 725.0 |
| 16 | 725.0 | 370.0 | 34 | 700.0 | 580.0 | 52 | 1740.0 | 245.0 |
| 17 | 145.0 | 665.0 | 35 | 685.0 | 595.0 | | | |
| 18 | 415.0 | 635.0 | 36 | 685.0 | 610.0 | | | |

## Bier127

TSP problem of TSPlib from Reinelt [117]: coordinates of 127 beer gardens in Augsburg, Germany; Euclidean distances; length of optimal tour $f(\mathbf{x}^*) = 118282$.

**Table B.3.** Coordinates of problem *bier127*, Euclidean distances

| city | x | y | city | x | y | city | x | y | city | x | y | city | x | y |
|---|---|---|---|---|---|---|---|---|---|---|---|---|---|---|
| 1 | 9860 | 14152 | 27 | 9512 | 12412 | 53 | 7888 | 16936 | 79 | 10556 | 11948 | 105 | 10440 | 14036 |
| 2 | 9396 | 14616 | 28 | 7772 | 11020 | 54 | 8236 | 15428 | 80 | 10324 | 11716 | 106 | 10672 | 13804 |
| 3 | 11252 | 14848 | 29 | 8352 | 10672 | 55 | 9512 | 17400 | 81 | 11484 | 9512 | 107 | 1160 | 18560 |
| 4 | 11020 | 13456 | 30 | 9164 | 12876 | 56 | 9164 | 16008 | 82 | 11484 | 7540 | 108 | 10788 | 13572 |
| 5 | 9512 | 15776 | 31 | 9744 | 12528 | 57 | 8700 | 15312 | 83 | 11020 | 7424 | 109 | 15660 | 11368 |
| 6 | 10788 | 13804 | 32 | 8352 | 10324 | 58 | 11716 | 16008 | 84 | 11484 | 9744 | 110 | 15544 | 12760 |
| 7 | 10208 | 14384 | 33 | 8236 | 11020 | 59 | 12992 | 14964 | 85 | 16936 | 12180 | 111 | 5336 | 18908 |
| 8 | 11600 | 13456 | 34 | 8468 | 12876 | 60 | 12412 | 14964 | 86 | 17052 | 12064 | 112 | 6264 | 19140 |
| 9 | 11252 | 14036 | 35 | 8700 | 14036 | 61 | 12296 | 15312 | 87 | 16936 | 11832 | 113 | 11832 | 17516 |
| 10 | 10672 | 15080 | 36 | 8932 | 13688 | 62 | 12528 | 15196 | 88 | 17052 | 11600 | 114 | 10672 | 14152 |
| 11 | 11136 | 14152 | 37 | 9048 | 13804 | 63 | 15312 | 6612 | 89 | 13804 | 18792 | 115 | 10208 | 15196 |
| 12 | 9860 | 13108 | 38 | 8468 | 12296 | 64 | 11716 | 16124 | 90 | 12064 | 14964 | 116 | 12180 | 14848 |
| 13 | 10092 | 14964 | 39 | 8352 | 12644 | 65 | 11600 | 19720 | 91 | 12180 | 15544 | 117 | 11020 | 10208 |
| 14 | 9512 | 13340 | 40 | 8236 | 13572 | 66 | 10324 | 17516 | 92 | 14152 | 18908 | 118 | 7656 | 17052 |
| 15 | 10556 | 13688 | 41 | 9164 | 13340 | 67 | 12412 | 13340 | 93 | 5104 | 14616 | 119 | 16240 | 8352 |
| 16 | 9628 | 14036 | 42 | 8004 | 12760 | 68 | 12876 | 12180 | 94 | 6496 | 17168 | 120 | 10440 | 14732 |
| 17 | 10904 | 13108 | 43 | 8584 | 13108 | 69 | 13688 | 10904 | 95 | 5684 | 13224 | 121 | 9164 | 15544 |
| 18 | 11368 | 12644 | 44 | 7772 | 14732 | 70 | 13688 | 11716 | 96 | 15660 | 10788 | 122 | 8004 | 11020 |
| 19 | 11252 | 13340 | 45 | 7540 | 15080 | 71 | 13688 | 12528 | 97 | 5336 | 10324 | 123 | 5684 | 11948 |
| 20 | 10672 | 13340 | 46 | 7424 | 17516 | 72 | 11484 | 13224 | 98 | 812 | 6264 | 124 | 9512 | 16472 |
| 21 | 11020 | 13108 | 47 | 8352 | 17052 | 73 | 12296 | 12760 | 99 | 14384 | 20184 | 125 | 13688 | 17516 |
| 22 | 11020 | 13340 | 48 | 7540 | 16820 | 74 | 12064 | 12528 | 100 | 11252 | 15776 | 126 | 11484 | 8468 |
| 23 | 11136 | 13572 | 49 | 7888 | 17168 | 75 | 12644 | 10556 | 101 | 9744 | 3132 | 127 | 3248 | 14152 |
| 24 | 11020 | 13688 | 50 | 9744 | 15196 | 76 | 11832 | 11252 | 102 | 10904 | 3480 | | | |
| 25 | 8468 | 11136 | 51 | 9164 | 14964 | 77 | 11368 | 12296 | 103 | 7308 | 14848 | | | |
| 26 | 8932 | 12064 | 52 | 9744 | 16240 | 78 | 11136 | 11020 | 104 | 16472 | 16472 | | | |

**Gr666**

TSP problem of TSPlib from Reinelt [117]: coordinates of 666 cities on earth; geographical distances; length of optimal tour $f(\mathbf{x}^*) = 294358$. We skip the presentation of the coordinates.

## B.2   SAT-Problem

**3-SAT**

Boolean / propositional satisfiability problem, each clause contains $k = 3$ literals; NP-complete; random generation of formulas with $N = 50$ variables and 150 clauses; fitness function:

$$\frac{\#\text{of true clauses}}{\#\text{of all clauses}} \tag{B.1}$$

For the experiments in chapter 6, fifty random formulas were generated according to the above specifications, and each formula was tested 10 times, i.e. the experimental results are based on 500 runs.

# References


1. Angeline, P.J.: Adaptive and self-adaptive evolutionary computations. In: Palaniswami, M., Attikiouzel, Y. (eds.) Computational Intelligence: A Dynamic Systems Perspective, pp. 152–163. IEEE Press, Los Alamitos (1995)
2. Arabas, J., Michalewicz, Z., Mulawka, J.J.: Gavaps - a genetic algorithm with varying population size. In: Proceedings of the International Conference on Evolutionary Computation, pp. 73–78 (1994)
3. Auger, A.: Convergence results for $(1, \lambda)$-sa-es using the theory of $\phi$-irreducible markov chains. In: Proceedings of the Evolutionary Algorithms workshop of the 30th International Colloquium on Automata, Languages and Programming (2003)
4. Bagley, J.D.: The behavior of adaptive systems which employ genetic and correlation algorithms. PhD thesis, University of Michigan (1967)
5. Bartz-Beielstein, T., Lasarczyk, C., Preuß, M.: Sequential parameter optimization. In: McKay, B., et al. (eds.) Proceedings of the IEEE Congress on Evolutionary Computation - CEC, vol. 1, pp. 773–780. IEEE Press, Los Alamitos (2005)
6. Bäck, T.: Self-adaptation in genetic algorithms. In: Proceedings of the 1st European Conference on Artificial Life - ECAL, pp. 263–271 (1991)
7. Bäck, T.: The interaction of mutation rate, selection, and self-adaptation within a genetic algorithm. In: Proceedings of the 2th Conference on Parallel Problem Solving from Nature - PPSN II, pp. 85–94 (1992)
8. Bäck, T., Fogel, D.B., Michalewicz, Z. (eds.): Handbook of Evolutionary Computation. IOP Publishing Ltd., Bristol (1997)
9. Bäck, T., Schutz, M.: Intelligent mutation rate control in canonical genetic algorithms. In: International Symposium on Methodologies for Intelligent Systems, pp. 158–167 (1996)
10. Bean, J.C.: Genetics and random keys for sequencing and optimization. Technical report, Department of Industrial and Operations Engineering, The University of Michigan (1992)
11. Bean, J.C., Hadj-Alouane, A.B.: A dual genetic algorithm for bounded integer programs. Technical report, University of Michigan (1992)
12. Belur, S.V.: CORE: Constrained optimization by random evolution. In: Koza, J.R. (ed.) Late Breaking Papers at the Genetic Programming 1997 Conference, Stanford University, California, Stanford Bookstore, July 1997, pp. 280–286 (1997)
13. Berlik, S.: A step size preserving directed mutation operator. In: Deb, K., et al. (eds.) GECCO 2004. LNCS, vol. 3103, pp. 786–787. Springer, Heidelberg (2004)

14. Beyer, H.-G.: An alternative explanation for the manner in which genetic algorithms operate. BioSystems 41(1), 1–15 (1997)
15. Beyer, H.-G.: On the performance of $(1, \lambda)$-evolution strategies for the ridge function class. IEEE Transactions on Evolutionary Computation 5(3), 218–235 (2001)
16. Beyer, H.-G.: The Theory of Evolution Strategies. Springer, Berlin (2001)
17. Beyer, H.-G., Meyer-Nieberg, S.: Self-adaptation on the ridge function class: First results for the sharp ridge. In: Proceedings of the 9th Conference on Parallel Problem Solving from Nature - PPSN IX, pp. 72–81 (2006)
18. Beyer, H.-G., Schwefel, H.-P.: Evolution strategies - A comprehensive introduction. Natural Computing 1, 3–52 (2002)
19. Bosman, P.A.N., Thierens, D.: Expanding from discrete to continuous estimation of distribution algorithms: The idea. In: Schwefel, H.-P., Schoenauer, M., Deb, K., Rudolph, G., Yao, X., Lutton, E., Merelo, J.J. (eds.) PPSN 2000. LNCS, vol. 1917, pp. 767–776. Springer, Heidelberg (2000)
20. Bosman, P.A.N., Thierens, D.: Adaptive variance scaling in continuous multi-objective estimation-of-distribution algorithms. In: Proceedings of the 9th conference on genetic and evolutionary computation - GECCO, pp. 500–507. ACM Press, New York (2007)
21. Broyden, C.G.: The convergence of a class of double-rank minimization algorithms. I: General considerations. J. Inst. Math. 6, 76–90 (1970)
22. Cai, Y., Sun, X., Xu, H., Jia, P.: Cross entropy and adaptive variance scaling in continuous eda. In: Proceedings of the 9th conference on genetic and evolutionary computation - GECCO, pp. 609–616. ACM Press, New York (2007)
23. Coello Coello, C.A.: Treating constraints as objectives for single-objective evolutionary optimization. Engineering Optimization 32(3), 275–308 (2000)
24. Coello Coello, C.A.: Use of a self-adaptive penalty approach for engineering optimization problems. Computers in Industry 41(2), 113–127 (2000)
25. Coello Coello, C.A.: Theoretical and numerical constraint handling techniques used with evolutionary algorithms: A survey of the state of the art. Computer Methods in Applied Mechanics and Engineering 191(11-12), 1245–1287 (2002)
26. Croes, G.: A method for solving traveling salesman problem. In: Operations Research, pp. 791–812 (1992)
27. Titterington, U.M.D., Smith, A.: Statistical Analysis of Finite Mixture Distributions. Wiley, Chichester (1986)
28. Darwin, C.: On the Origin of Species. John Murray, London (1859)
29. Davidon, W.: Variable metric methods for minimization. SIAM Journal Optimization 1, 1–17 (1991)
30. Davis, L.: Adapting operator probabilities in genetic algorithms. In: Proceedings of the 3rd international conference on Genetic algorithms, pp. 61–69. Morgan Kaufmann Publishers Inc., San Francisco (1989)
31. de Landgraaf, W., Eiben, A., Nannen, V.: Parameter calibration using meta-algorithms. In: Proceedings of the IEEE Congress on Evolutionary Computation - CEC, pp. 71–78 (2007)
32. Deb, K.: An efficient constraint handling method for genetic algorithms. Computer Methods in Applied Mechanics and Engineering (2001)
33. Deb, K., Beyer, H.-G.: Self-adaptation in real-parameter genetic algorithms with simulated binary crossover. In: Banzhaf, W., Daida, J., Eiben, A.E., Garzon, M.H., Honavar, V., Jakiela, M., Smith, R.E. (eds.) Proceedings of the Genetic and Evolutionary Computation Conference, vol. 1, pp. 172–179. Morgan Kaufmann Publishers Inc., San Francisco (1999)

34. Deb, K., Georg Beyer, H.: Self-adaptive genetic algorithms with simulated binary crossover. Evolutionary Computation 9(2), 197–221 (2001)
35. Eiben, A., Schut, M.C., de Wilde, A.: Is self-adaptation of selection pressure and population size possible? - a case study. In: Proceedings of the 9th Conference on Parallel Problem Solving from Nature - PPSN IX, pp. 900–909 (2006)
36. Eiben, A.E., Aarts, E.H.L., van Hee, K.M.: Global convergence of genetic algorithms: A markov chain analysis. In: PPSN I: Proceedings of the 1st Workshop on Parallel Problem Solving from Nature, pp. 4–12. Springer, Berlin (1991)
37. Eiben, A.E., Hinterding, R., Michalewicz, Z.: Parameter control in evolutionary algorithms. IEEE Transactions on Evolutionary Computation 3(2), 124–141 (1999)
38. Eiben, A.E., Smith, J.E.: Introduction to Evolutionary Computing. Springer, Berlin (2003)
39. Eshelman, L.J., Schaffer, J.D.: Real-coded genetic algorithms and interval-schemata. In: Proceedings of the FOGA, pp. 187–202 (1992)
40. Fiacco, A., McCormick, G.: The sequential unconstrained minimization technique for nonlinear programming - a primal-dual method. Mgmt. Sci. 10, 360–366 (1964)
41. Fogarty, T.C.: Varying the probability of mutation in the genetic algorithm. In: Proceedings of the 3rd International Conference on Genetic Algorithms, pp. 104–109. Morgan Kaufmann, San Francisco (1989)
42. Fogel, D.B.: Evolving Artificial Intelligence. PhD thesis, University of California, San Diego (1992)
43. Fogel, D.B., Fogel, L.J., Atma, J.W.: Meta-evolutionary programming. In: Proceedings of 25th Asilomar Conference on Signals, Systems & Computers, pp. 540–545 (1991)
44. Fogel, L.J., Owens, A.J., Walsh, M.J.: Artificial Intelligence through Simulated Evolution. Wiley, New York (1966)
45. Garey, M.R., Johnson, D.S.: Computers and Intractability; A Guide to the Theory of NP-Completeness. W. H. Freeman & Co., New York (1990)
46. Glickman, M., Sycara, K.: Reasons for premature convergence of self-adapting mutation rates. In: Proceedings of the Congress on Evolutionary Computation, vol. 1, pp. 62–69 (July 2000)
47. Goldberg, D.: Genetic Algorithms in Search, Optimization and Machine Learning. Addison-Wesley, Reading (1989)
48. Goldberg, D.E., Lingle, R.: Alleles, loci and the traveling salesman problem. In: Grefenstette (ed.) Proceedings of the 1st International Conference on Genetic Algorithms and Their Applications, pp. 154–159 (1985)
49. Grahl, J., Bosman, P.A.N., Minner, S.: Convergence phases, variance trajectories, and runtime analysis of continuous edas. In: Proceedings of the 9th conference on genetic and evolutionary computation - GECCO, pp. 516–522. ACM Press, New York (2007)
50. Grahl, J., Minner, S., Rothlauf, F.: Behaviour of umdac with truncation selection on monotonous functions. In: Proceedings of the IEEE Congress on Evolutionary Computation - CEC, vol. 3, pp. 2553–2559 (2005)
51. Grefenstette, J.: Optimization of control parameters for genetic algorithms. IEEE Trans. Syst. Man Cybern. 16(1), 122–128 (1986)
52. Hansen, N.: The cma evolution strategy: A tutorial. Technical report, TU Berlin, ETH Zürich (2005)
53. Hansen, N.: An analysis of mutative sigma self-adaptation on linear fitness functions. Evolutionary Computation 14(3), 255–275 (2006)

54. Hansen, N., Ostermeier, A.: Completely derandomized self-adaptation in evolution strategies. Evolutionary Computation 9(2), 159–195 (2001)
55. Hildebrand, L.: Asymmetrische Evolutionsstrategien. PhD thesis, University of Dortmund (2002)
56. Hinterding, R., Michalewicz, Z., Peachey, T.C.: Self-adaptive genetic algorithm for numeric functions. In: Ebeling, W., Rechenberg, I., Voigt, H.-M., Schwefel, H.-P. (eds.) PPSN 1996. LNCS, vol. 1141, pp. 420–429. Springer, Berlin (1996)
57. Hoffmeister, F., Sprave, J.: Problem-independent handling of constraints by use of metric penalty functions. In: Fogel, L.J., Angeline, P.J., Bäck, T. (eds.) Proceedings of the 5th Conference on Evolutionary Programming - EP 1996, pp. 289–294. MIT Press, Cambridge (1996)
58. Holland, J.H.: Adaptation in Natural and Artificial Systems. University of Michigan Press, Ann Arbor (1975)
59. Holland, J.H.: Adaptation in natural and artificial systems, 1st edn., 1975. MIT Press, Cambridge (1992)
60. Homaifar, A., Lai, S.H.Y., Qi, X.: Constrained Optimization via Genetic Algorithms. Simulation 62(4), 242–254 (1994)
61. Igel, C., Toussaint, M.: Neutrality and self-adaptation. Natural Computing 2(2), 117–132 (2003)
62. Jansen, T., Wegener, I.: Real royal road functions - where crossover provably is essential. In: Spector, L., Goodman, E.D., Wu, A., Langdon, W., Voigt, H.-M., Gen, M., Sen, S., Dorigo, M., Pezeshk, S., Garzon, M.H., Burke, E. (eds.) Proceedings of the 3rd conference on genetic and evolutionary computation - GECCO, pp. 375–382. Morgan Kaufmann, San Francisco (2001)
63. Jägersküpper, J.: Rigorous runtime analysis of the (1+1) es: 1/5-rule and ellipsoidal fitness landscapes. In: Wright, A.H., Vose, M.D., De Jong, K.A., Schmitt, L.M. (eds.) FOGA 2005. LNCS, vol. 3469, pp. 260–281. Springer, Heidelberg (2005)
64. Jägersküpper, J.: Probabilistic runtime analysis of $(1 + \lambda)$es using isotropic mutations. In: Proceedings of the 8th conference on Genetic and evolutionary computation - GECCO, pp. 461–468. ACM Press, New York (2006)
65. Jimenez, F., Verdegay, J.L.: Evolutionary techniques for constrained optimization problems. In: Zimmermann, H.-J. (ed.) 7th European Congress on Intelligent Techniques and Soft Computing (EUFIT 1999), Aachen, Verlag Mainz (1999)
66. Joines, J., Houck, C.: On the use of non-stationary penalty functions to solve nonlinear constrained optimization problems with GAs. In: Fogel, D.B. (ed.) Proceedings of the 1st IEEE Conference on Evolutionary Computation, Orlando, Florida, pp. 579–584. IEEE Press, Los Alamitos (1994)
67. Jong, K.A.D.: An analysis of the behavior of a class of genetic adaptive systems. PhD thesis, University of Michigan (1975)
68. Kanji, G.: 100 Statistical Tests. SAGE Publications, London (1993)
69. Kennedy, J., Eberhart, R.: Particle swarm optimization. In: Proceedings of IEEE International Conference on Neural Networks, pp. 1942–1948 (1995)
70. Koza, J.R.: Genetic Programming: On the Programming of Computers by Means of Natural Selection. MIT Press, Cambridge (1992)
71. Koziel, S., Michalewicz, Z.: Evolutionary algorithms, homomorphous mappings, and constrained parameter optimization. Evolutionary Computation 7(1), 19–44 (1999)
72. Kramer, O., Brügger, S., Lazovic, D.: Sex and death: towards biologically inspired heuristics for constraint handling. In: Proceedings of the 9th conference on genetic and evolutionary computation - GECCO, pp. 666–673. ACM Press, New York (2007)

73. Kramer, O., Gloger, B., Goebels, A.: An experimental analysis of evolution strategies and particle swarm optimisers using design of experiments. In: Proceedings of the 9th conference on genetic and evolutionary computation - GECCO, pp. 674–681. ACM Press, New York (2007)

74. Kramer, O., Koch, P.: Self-adaptive partially mapped crossover. In: Proceedings of the 9th conference on genetic and evolutionary computation - GECCO, p. 1523. ACM Press, New York (2007)

75. Kramer, O., Schwefel, H.-P.: On three new approaches to handle constraints within evolution strategies. Natural Computing 5(4), 363–385 (2006)

76. Kramer, O., Stein, B., Wall, J.: Ai and music: Toward a taxonomy of problem classes. In: ECAI, pp. 695–696 (2006)

77. Kramer, O., Ting, C.-K., Büning, H.K.: A mutation operator for evolution strategies to handle constrained problems. In: Proceedings of the 7th conference on genetic and evolutionary computation - GECCO, pp. 917–918 (2005)

78. Kramer, O., Ting, C.-K., Büning, H.K.: A new mutation operator for evolution strategies for constrained problems. In: Proceedings of the IEEE Congress on Evolutionary Computation - CEC, pp. 2600–2606 (2005)

79. Kuri-Morales, A., Quezada, C.V.: A universal eclectic genetic algorithm for constrained optimization. In: Proceedings 6th European Congress on Intelligent Techniques & Soft Computing, EUFIT 1998, Aachen, September 1998, pp. 518–522. Verlag Mainz (1998)

80. Kursawe, F.: Grundlegende empirische Untersuchungen der Parameter von Evolutionsstrategien - Metastrategien. PhD thesis, University of Dortmund (1999)

81. Larranaga, P., Lozano, J.: Estimation of Distribution Algorithms. A New Tool for Evolutionary Computation. Kluwer Academic Publishers, Dordrecht (2001)

82. Lewis, R., Torczon, V., Trosset, M.: Direct search methods: Then and now. Journal of Computational and Applied Mathematics 124(1-2), 191–207 (2000)

83. Liang, J., Runarsson, T.P., Mezura-Montes, E., Clerc, M., Suganthan, P., Coello Coello, C.A., Deb, K.: Problem definitions and evaluation criteria for the CEC 2006, special session on constrained real-parameter optimization. In Technical Report, Singapore, Nanyang Technological University (2006)

84. Liang, J., Suganthan, P.: Dynamic multi-swarm particle swarm optimizer with a novel constraint-handling mechanism. In: Yen, G.G., Lucas, S.M., Fogel, G., Kendall, G., Salomon, R., Zhang, B.-T., Coello Coello, C.A., Runarsson, T.P. (eds.) Proceedings of the 2006 IEEE Congress on Evolutionary Computation, Vancouver, July 2006, pp. 9–16. IEEE Press, Los Alamitos (2006)

85. Liang, K.-H., Yao, X., Liu, Y., Newton, C.S., Hoffman, D.: An experimental investigation of self-adaptation in evolutionary programming. In: Porto, V.W., Waagen, D. (eds.) EP 1998. LNCS, vol. 1447, pp. 291–300. Springer, Heidelberg (1998)

86. Lozano, J.A., Larranaga, P., Inza, I.: Towards a New Evolutionary Computation: Advances on Estimation of Distribution Algorithms. Studies in Fuzziness and Soft Computing. Springer, Berlin (2006)

87. Lukacs, E.: Stochastic Convergence, 2nd edn. Academic Press, New York (1975)

88. Maruo, M.H., Lopes, H.S., Delgado, M.R.: Self-adapting evolutionary parameters: Encoding aspects for combinatorial optimization problems. In: Raidl, G.R., Gottlieb, J. (eds.) EvoCOP 2005. LNCS, vol. 3448, pp. 154–165. Springer, Heidelberg (2005)

89. Meer, K.: Simulated annealing versus metropolis for a tsp instance. Inf. Process. Lett. 104(6), 216–219 (2007)

90. Mercer, R.E., Sampson, J.R.: Adaptive search using a reproductive metaplan. Kybernetes 7, 215–228 (1978)
91. Meyer-Nieberg, S., Beyer, H.-G.: Self-adaptation in evolutionary algorithms. In: Lobo, F.G., Lima, C.F., Michalewicz, Z. (eds.) Parameter Setting in Evolutionary Algorithms. Springer, Berlin (2007)
92. Mezura-Montes, E., Velazquez-Reyes, J., Coello Coello, C.A.: Modified differential evolution for constrained optimization. In: Yen, G.G., Lucas, S.M., Fogel, G., Kendall, G., Salomon, R., Zhang, B.-T., Coello Coello, C.A., Runarsson, T.P. (eds.) Proceedings of the 2006 IEEE Congress on Evolutionary Computation, Vancouver, July 2006, pp. 25–32. IEEE Press, Los Alamitos (2006)
93. Mühlenbein, H.: How genetic algorithms really work: Mutation and hillclimbing. In: Proceedings of the 2th Conference on Parallel Problem Solving from Nature - PPSN II, pp. 15–26 (1992)
94. Michalewicz, Fogel, D.B.: How to Solve It: Modern Heuristics. Springer, Berlin (2000)
95. Mitchell, M., Holland, J.H., Forrest, S.: When will a genetic algorithm outperform hill climbing. In: Cowan, J.D., Tesauro, G., Alspector, J. (eds.) Advances in Neural Information Processing Systems, vol. 6, pp. 51–58. Morgan Kaufmann Publishers, Inc., San Francisco (1994)
96. Montes, E.M., Coello Coello, C.A.: A simple multi-membered evolution strategy to solve constrained optimization problems. IEEE Transactions on Evolutionary Computation 9(1), 1–17 (2005)
97. Nannen, V., Eiben, A.: A method for parameter calibration and relevance estimation in evolutionary algorithms. In: Proceedings of the 8th conference on Genetic and evolutionary computation - GECCO, pp. 183–190. ACM Press, New York (2006)
98. Nannen, V., Eiben, A.: Relevance estimation and value calibration of evolutionary algorithm parameters. In: IJCAI, pp. 975–980 (2007)
99. Nelder, J., Mead, R.: A simplex method for function minimization. The Computer Journal 7, 308–313 (1964)
100. Ocenasek, J., Kern, S., Hansen, N., Koumoutsakos, P.: A mixed bayesian optimization algorithm with variance adaptation. In: Yao, X., Burke, E.K., Lozano, J.A., Smith, J., Merelo-Guervós, J.J., Bullinaria, J.A., Rowe, J.E., Tiňo, P., Kabán, A., Schwefel, H.-P. (eds.) PPSN 2004. LNCS, vol. 3242, pp. 352–361. Springer, Heidelberg (2004)
101. Ostermeier, A., Gawelczyk, A., Hansen, N.: A derandomized approach to self adaptation of evolution strategies. Evolutionary Computation 2(4), 369–380 (1994)
102. Ostermeier, A., Gawelczyk, A., Hansen, N.: A derandomized approach to self adaptation of evolution strategies. Evolutionary Computation 2(4), 369–380 (1995)
103. Paredis, J.: Co-evolutionary constraint satisfaction. In: Davidor, Y., Männer, R., Schwefel, H.-P. (eds.) PPSN 1994. LNCS, vol. 866, pp. 46–55. Springer, Heidelberg (1994)
104. Parmee, I.C., Purchase, G.: The development of a directed genetic search technique for heavily constrained design spaces. In: Parmee, I.C. (ed.) Proceedings of the Conference on Adaptive Computing in Engineering Design and Control 1994, pp. 97–102. University of Plymouth, Plymouth (1994)
105. Pelikan, M., Goldberg, D.E., Tsutsui, S.: Getting the best of both worlds: discrete and continuous genetic and evolutionary algorithms in concert. Inf. Sci. 156(3-4), 147–171 (2003)

106. Pelikan, M., Sastry, K., Cantu-Paz, E.: Scalable Optimization via Probabilistic Modeling: From Algorithms to Applications. Studies in Computational Intelligence. Springer, Berlin (2006)
107. Pennisi, E.: Working the (gene count) numbers: Finally, a firm answer? Science 316(5828), 1113 (2007)
108. Poli, R., Chio, C.D., Langdon, W.B.: Exploring extended particle swarms: a genetic programming approach. In: Proceedings of the 7th conference on genetic and evolutionary computation - GECCO, pp. 169–176 (2005)
109. Powell, D., Skolnick, M.M.: Using genetic algorithms in engineering design optimization with non-linear constraints. In: Forrest, S. (ed.) Proceedings of the 5th International Conference on Genetic Algorithms - ICGA 1993, University of Illinois at Urbana-Champaign, July 1993, pp. 424–431. Morgan Kaufmann, San Francisco (1993)
110. Powell, M.: An efficient method for finding the minimum of a function of several variables without calculating derivatives. The Computer Journal 7(2), 155–162 (1964)
111. Priesterjahn, S., Kramer, O., Weimer, A., Goebels, A.: Evolution of reactive rules in multi player computer games based on imitation. In: Wang, L., Chen, K., S. Ong, Y. (eds.) ICNC 2005. LNCS, vol. 3612, pp. 744–755. Springer, Heidelberg (2005)
112. Priesterjahn, S., Kramer, O., Weimer, A., Goebels, A.: Evolution of human-competitive agents in modern computer games. In: Yen, G.G., Lucas, S.M., Fogel, G., Kendall, G., Salomon, R., Zhang, B.-T., Coello Coello, C.A., Runarsson, T.P. (eds.) Proceedings of the 2006 IEEE Congress on Evolutionary Computation, Vancouver, July 2006, pp. 777–784. IEEE Press, Los Alamitos (2006)
113. Rechenberg, I.: Cybernetic solution path of an experimental problem. In: Ministry of Aviation, UK, Royal Aircraft Establishment (1965)
114. Rechenberg, I.: Evolutionsstrategie: Optimierung technischer Systeme nach Prinzipien der biologischen Evolution. Frommann-Holzboog, Stuttgart (1973)
115. Rechenberg, I.: Evolutionsstrategie 1994. Frommann-Holzboog, Stuttgart (1994)
116. Reed, J., Toombs, R., Barricelli, N.A.: Simulation of biological evolution and machine learning: I. selection of self-reproducing numeric patterns by data processing machines, effects of hereditary control, mutation type and crossing. Journal of Theoretical Biology 17, 319–342 (1967)
117. Reinelt, G.: Tsplib - a traveling salesman problem library. ORSA J. Comput. 3, 376–384 (1991)
118. Riche, R.G.L., Knopf-Lenoir, C., Haftka, R.T.: A segregated genetic algorithm for constrained structural optimization. In: Eshelman, L.J. (ed.) Proceedings of the 6th International Conference on Genetic Algorithms - ICGA 1995, University of Pittsburgh, July 1995, pp. 558–565. Morgan Kaufmann, San Francisco (1995)
119. Rosenberg, R.S.: Simulation of genetic populations with biochemical properties. PhD thesis, University of Michigan (1967)
120. Rosenblatt, F.: The perceptron. a probabilistic model for information storage and organization in the brain. Psychological Reviews 65, 386–408 (1958)
121. Rosenbrock, H.: An automatic method for finding the greatest or least value of a function. The Computer Journal 3(3), 175–184 (1960)
122. Rudlof, S., Köppen, M.: Stochastic hill climbing by vectors of normal distributions. In: Proceedings of the 1st Online Workshop on Soft Computing, Nagoya, Japan (1996)
123. Rudolph, G.: Massively parallel simulated annealing and its relation to evolutionary algorithms. Evolutionary Computation 1(4), 361–383 (1993)

124. Rudolph, G.: Convergence of evolutionary algorithms in general search spaces. In: Proceedings of the Proceedings of the International Conference on Evolutionary Computation, pp. 50–54 (1996)

125. Rudolph, G.: Finite Markov chain results in evolutionary computation: A tour d'horizon. Fundamenta Informaticae 35(1-4), 67–89 (1998)

126. Rudolph, G.: Self-adaptive mutations lead to premature convergence. IEEE Transactions on Evolutionary Computation 5(4), 410–414 (2001)

127. Schaffer, J.D., Caruana, R., Eshelman, L.J., Das, R.: A study of control parameters affecting online performance of genetic algorithms for function optimization. In: Proceedings of the 3rd International Conference on Genetic Algorithms - ICGA 1989, pp. 51–60 (1989)

128. Schaffer, J.D., Morishima, A.: An adaptive crossover distribution mechanism for genetic algorithms. In: Proceedings of the 2nd International Conference on Genetic Algorithms on Genetic algorithms and their application, pp. 36–40. Lawrence Erlbaum Associates, Inc., Mahwah (1987)

129. Schoenauer, M., Michalewicz, Z.: Evolutionary computation at the edge of feasibility. In: Voigt, H.-M., Ebeling, W., Rechenberg, I., Schwefel, H.-P. (eds.) PPSN 1996. LNCS, vol. 1141, pp. 245–254. Springer, Berlin (1996)

130. Schoenauer, M., Xanthakis, S.: Constrained GA optimization. In: Forrest, S. (ed.) Proceedings of the 5th International Conference on Genetic Algorithms - ICGA 1993, pp. 573–580. Morgan Kaufman Publishers Inc., San Francisco (1993)

131. Schwefel, H.-P.: Kybernetische Evolution als Strategie der experimentellen Forschung in der Strömungstechnik. diploma thesis, TU Berlin, Hermann Föttinger-Institut für Strömungstechnik (March 1965)

132. Schwefel, H.-P.: Adaptive Mechanismen in der biologischen Evolution und ihr Einfluss auf die Evolutionsgeschwindigkeit. Interner Bericht der Arbeitsgruppe Bionik und Evolutionstechnik am Institut für Mess- und Regelungstechnik, TU Berlin (July 1974)

133. Schwefel, H.-P.: Evolutionsstrategie und numerische Optimierung. PhD thesis, TU Berlin (1975)

134. Schwefel, H.-P.: Numerische Optimierung von Computer-Modellen mittels der Evolutions strategie. Birkhäuser, Basel (1977)

135. Schwefel, H.-P.: Evolution and Optimum Seeking. In: Sixth-Generation Computer Technology. Wiley Interscience, New York (1995)

136. Sebag, M., Ducoulombier, A.: Extending population-based incremental learning to continuous search spaces. In: Eiben, A.E., Bäck, T., Schoenauer, M., Schwefel, H.-P. (eds.) PPSN 1998. LNCS, vol. 1498, pp. 418–427. Springer, Heidelberg (1998)

137. Semenov, M.A., Terkel, D.A.: Analysis of convergence of an evolutionary algorithm with self-adaptation using a stochastic lyapunov function. Evol. Comput. 11(4), 363–379 (2003)

138. Shapiro, J., Prugel-Bennett, A., Rattray, M.: A statistical mechanical formulation of the dynamics of genetic algorithms. In: Evolutionary Computing, AISB Workshop, pp. 17–27 (1994)

139. Shi, Y., Eberhart, R.: A modified particle swarm optimizer. In: Proceedings of the International Conference on Evolutionary Computation, pp. 69–73 (1998)

140. Sinha, A., Srinivasan, A., Deb, K.: A population-based, parent centric procedure for constrained real-parameter optimization. In: Yen, G.G., Lucas, S.M., Fogel, G., Kendall, G., Salomon, R., Zhang, B.-T., Coello Coello, C.A., Runarsson, T.P. (eds.) Proceedings of the 2006 IEEE Congress on Evolutionary Computation, Vancouver, pp. 239–245. IEEE Press, Los Alamitos (2006)

141. Skinner, C., Riddle, P.J., Triggs, C.: Mathematics prevents bloat. In: Proceedings of the IEEE Congress on Evolutionary Computation - CEC, pp. 390–395 (2005)
142. Smith, J.: Modelling gas with self adaptive mutation rates. In: Proceedings of the Genetic and Evolutionary Computation Conference, pp. 599–606 (2001)
143. Smith, J., Fogarty, T.C.: Recombination strategy adaptation via evolution of gene linkage. In: Proceedings of the IEEE Congress on Evolutionary Computation - CEC, pp. 826–831 (1996)
144. Smith, J., Fogarty, T.C.: Self adaptation of mutation rates in a steady state genetic algorithm. In: Proceedings of the International Conference on Evolutionary Computation, pp. 318–323 (1996)
145. Smith, R.E.: Adaptively resizing populations: An algorithm and analysis. In: Proceedings of the 5th International Conference on Genetic Algorithms, p. 653. Morgan Kaufmann Publishers Inc., San Francisco (1993)
146. Snyder, L., Daskin, M.: A random-key genetic algorithm for the generalized traveling salesman problem. Technical report, Department of Industrial Engineering and Management Sciences, Northwestern University (2001)
147. Spears, W.M.: Adapting crossover in evolutionary algorithms. In: McDonnell, J.R., Reynolds, R.G., Fogel, D.B. (eds.) Proceedings of the Fourth Annual Conference on Evolutionary Programming, pp. 367–384. MIT Press, Cambridge (1995)
148. Stone, C., Smith, J.: Strategy parameter variety in self-adaptation of mutation rates. In: Proceedings of the Genetic and Evolutionary Computation Conference - GECCO, pp. 586–593. Morgan Kaufmann Publishers Inc., San Francisco (2002)
149. Suganthan, P.N., Hansen, N., Liang, J.J., Deb, K., Chen, Y.P., Auger, A., Tiwari, S.: Problem definitions and evaluation criteria for the CEC 2005 special session on real-parameter optimization. Technical report, Nanyang Technological University (May 2005)
150. Surry, P.D., Radcliffe, N.J., Boyd, I.D.: A multi-objective approach to constrained optimisation of gas supply networks: The COMOGA Method. In: Fogarty, T.C. (ed.) Evolutionary Computing. AISB Workshop. Selected Papers. LNCS, pp. 166–180. Springer, Berlin (1995)
151. Syswerda, G.: Simulated crossover in genetic algorithms. In: Proceedings of the FOGA, pp. 239–255 (1992)
152. Takahama, T., Sakai, S.: Constrained optimization by the epsilon constrained differential evolution with gradient-based mutation and feasible elites. In: Yen, G.G., Lucas, S.M., Fogel, G., Kendall, G., Salomon, R., Zhang, B.-T., Coello Coello, C.A., Runarsson, T.P. (eds.) Proceedings of the 2006 IEEE Congress on Evolutionary Computation, Vancouver, pp. 1–8. IEEE Press, Los Alamitos (2006)
153. Ting, C.-K.: An analysis of the effectiveness of multi-parent crossover. In: Yao, X., Burke, E.K., Lozano, J.A., Smith, J., Merelo-Guervós, J.J., Bullinaria, J.A., Rowe, J.E., Tiňo, P., Kabán, A., Schwefel, H.-P. (eds.) PPSN 2004. LNCS, vol. 3242, pp. 130–139. Springer, Berlin (2004)
154. Tomassini, M. (ed.): Spatially Structured Evolutionary Algorithms: Artificial Evolution in Space and Time. Springer, Heidelberg (2005)
155. Vose, M.: The simple genetic algorithm. MIT Press, Cambridge (1999)
156. Weaver, R.F.: Molecular Biology. McGraw-Hill Higher Education, New York (2005)
157. Weinberg, R.: Computer simulation of a living cell. PhD thesis, University of Michigan (1970)

158. Whitley, D., Mathias, K., Fitzhorn, P.: Delta coding: An iterative search strategy for genetic algorithms. In: Belew, R., Booker, L. (eds.) Proceedings of the Fourth International Conference on Genetic Algorithms, pp. 77–84. Morgan Kaufman Publishers Inc., San Francisco (1991)
159. Whitley, D.L., Yoo, N.: Modeling simple genetic algorithms for permutation problems. In: Whitley, L.D., Vose, M.D. (eds.) Foundations of Genetic Algorithms III, pp. 163–184. Morgan Kaufmann Publishers Inc., San Francisco (1995)
160. Witt, C.: Runtime analysis of the $(\mu+1)$ EA on simple pseudo boolean functions. Evolutionary Computation 14(1), 65–86 (2006)
161. Wolpert, D.H., Macready, W.G.: No free lunch theorems for search. Technical report, Santa Fe, NM (1995)
162. Yao, X., Liu, Y.: Fast evolution strategies. In: Angeline, P.J., Reynolds, R.G., McDonnell, J.R., Eberhart, R. (eds.) Evolutionary Programming VI, pp. 151–161. Springer, Berlin (1997)
163. Yao, X., Liu, Y., Liang, K.-H., Lin, G.: Fast evolutionary algorithms. Springer, Berlin (2003)
164. Yuan, B., Gallagher, M.: On the importance of diversity maintenance in estimation of distribution algorithms. In: Proceedings of the 7th conference on genetic and evolutionary computation - GECCO, pp. 719–726. ACM Press, New York (2005)

# List of Figures

# List of Tables

# Index